EXTINCTION

DOUGLAS H. ERWIN

EXTIN

CTION

How Life on Earth Nearly
Ended 250 Million Years Ago

PRINCETON UNIVERSITY PRESS

Princeton and Oxford

Copyright © 2006 by Princeton University Press
Published by Princeton University Press, 41 William Street,
Princeton, New Jersey 08540
In the United Kingdom: Princeton University Press,
3 Market Place, Woodstock, Oxfordshire OX20 1SY

Library of Congress Cataloging-in-Publication Data

Erwin, Douglas H., 1958–
Extinction : how life on earth nearly ended
250 million years ago / Douglas H. Erwin.
p.   cm.
Includes bibliographical references and index.
ISBN-13: 978-0-691-00524-9 (cloth : alk. paper)
ISBN-10: 0-691-00524-9 (cloth : alk. paper)
1. Extinction (Biology)—History. 2. Catastrophes (Geology) 3. Geology,
Stratigraphic—Permian. 4. Paleontology, Stratigraphic. I. Title.
QE721.2.E97E96546 2005
576.8′4—dc22   2005043376

British Library Cataloging-in-Publication Data is available

This book has been composed in New Baskerville

Printed on acid-free paper. ∞

pup.princeton.edu

Printed in the United States of America

1   3   5   7   9   10   8   6   4   2

Contents

Acknowledgments

I owe a great debt to the many colleagues with whom I have worked on the problems of the end-Permian mass extinction, and without whose research, insights, and questions this book would not have been possible. Chief among them are my two longtime research collaborators, Sam Bowring from MIT, and Professor Jin Yugan of the Nanjing Institute of Geology and Palaeontology. Members of their groups have also made important contributions, particularly Mark Martin and Kathy Davidek at MIT and Zhu Zhuli, Wang Wei, and Wang Yue at Nanjing. We each have brought our own training and biases to this project, and together we may have actually made some progress. I am also greatly indebted to my longtime research assistant, Elisabeth Valiulis, without whose efforts in research support, design of graphics and illustrations, and other areas have made all of this research possible. Liz is responsible for most of the illustrations in this book, for which I am enormously grateful. For insights into the vertebrate extinction and hospitality in the Karoo of South Africa, I thank Bruce Rubidge of the Bernard Price Institute and his brother Robert and Robert's wife Marianne at Wellwood in the Karoo. My colleagues at the Department of Paleobiology at NMNH have been an invaluable source of support, particularly Kay Behrensmeyer, Bill DiMichele, and Conrad Labandeira. My colleagues in the Harvard/

MIT node of NASA's Astrobiology Institute have also been a great source of insight, particularly Andy Knoll, John Grotzinger, Dan Schrag, Charles Marshall, and John Hayes. Other significant contributors include Richard Bambach, Linda Elkins-Tanton, the late Bill Holser, Tong Jinnan, Arnie Miller, Chris Sidor, Roger Smith, Sherman Suter, Peter Ward, Bruce Wardlaw, Paul Wignall, Heather Wilson, and Rachel Wood.

I am particularly indebted to Richard Bambach for his perspicacious and detailed comments on an earlier draft of the entire manuscript, and to Paul Wignall for his helpful comments and corrections. I also deeply appreciate the comments of those who read various chapters of the manuscript, including Hallie Sims, Sam Bowring, Cindy Looy and Bruce Rubidge. Photographs were taken by the author except where indicated; illustrations were prepared by Elisabeth Valiulis.

For support of my research on the end-Permian mass extinction I gratefully acknowledge the Exobiology Program of NASA and the NASA Astrobiology Institute. The Smithsonian Institution and the National Museum of Natural History provided the freedom for me to pursue this work. Much of the first portion of this book was written during the tenure of an Overseas Visiting Fellowship at St. John's College, Cambridge, and I am indebted to the Master and Fellows of St. John's, to the Department of Earth Sciences, and particularly to Professor Simon Conway Morris and to Rachel Wood for their collegiality during my stay in Cambridge. More recently, the Santa Fe Institute has provided a congenial and intellectually exciting home away from home and in particular has supported my work on recoveries from mass extinctions and evolutionary innovations (chapters 9 and 10) through funding from the Thaw Charitable Trust.

Sam Elworthy patiently shepherded this project through to completion, and I greatly appreciate his perceptive comments, and Marsha Kunin's excellent copyediting. Finally, my deepest thanks to Wendy Wiswall for her love, support, and affection during the completion of this project.

EXTINCTION

CHAPTER 1
Introduction

San Rafael Swell, Near Green River, Utah

Early Triassic rocks are boring. It doesn't matter where you are, China, Europe or here in Utah; there is a certain similarity to them, and a dreadful monotony. A kind of austere beauty, but monotonous nonetheless. This great swath of sandstones and limestones flows across hundreds of square miles from Nevada up through Utah into Idaho and Montana, remnants of a shallow sea 250 million years old. In Idaho and Montana these rocks are buried by rock debris along the slopes of wonderful mountain valleys, or pine forests swallow them up (plants are no friend to the field geologist, except at lunch when you need some shade).

In Utah the Triassic is laid out for anyone who cares to look, although few do. The dry mesas of the San Rafael Swell west of Green River are home to rattlesnakes and lizards and visited by archeologists seeking the rock art of the ancient Fremont culture. Fossils are few and far between: a few species of fossil clams, the occasional snail or two, and the odd coiled ammonite, a distant relative of today's nautilus. Today the starfish, sea anemones, crabs, and snails in a tide pool at Santa Barbara, California, are very different from those in Baja California, much less Maine or Japan. These fossils in Utah are almost identical to those in rocks of the same age in northern Italy, Iran, and southern China. Farther north, clams are incredibly abundant: great pavements of the scalloplike clam *Claraia* cover the surface in their thousands or tens of thousands. Not dinosaurs perhaps, but to me far more valuable. Occasionally a few narrow bands of rock yield a richer trove of fossils. Rich is relative, of course, for there may be only nine or ten species of snails, but they pose an enigma far more compelling than the end of the dinosaurs.

To understand this enigma we need to travel farther back in time, to rocks 20 million years older in the Guadalupe Mountains of west Texas (figure 1.1). Instead of the scattered species of the Early Triassic of Utah, hundreds of species of snails have been found here. But these were dwarfed by a myriad of other marine animals. The Guadalupe Mountains stretch from Carlsbad Caverns southwestward to Guadalupe Mountains National Park. This 265-million-year-old fossilized reef once flourished across west Texas along the margins of what was then a warm, nourishing equatorial sea. The reef rivaled today's Great Barrier Reef of Australia in size and biotic diversity.

Today the canyons and caves of the Guadalupe Mountains dissect the reef for all to explore. There are few places in the world, of any age, where one can explore the inside of a reef, see how it was formed, and how it evolved over millions of years. Not surprisingly, geologists have been making pilgrimages to the Guadalupe Mountains for decades because erosion has exhumed the reef intact. At the base of the steep escarpment at McKittrick Creek one is standing on the ancient sea bottom of the Permian Basin looking up toward the reef some 1,200 feet above, just as one could today off the Bahamas or some other modern reef if all the water was removed. Hiking up the Permian Reef Trail in McKittrick Canyon is just like walking (or better, swimming) up the face of the reef as it was millions of years ago. The trail twists between juniper trees and across fine-grained limestones, then, climbing a bit more steeply, the trail dodges huge blocks of limestone. Turning to look across to the opposite side of the canyon, hikers can see the outline of avalanches of the reef into deep water clearly preserved. One huge block of limestone, a chunk of reef two hundred feet long and forty feet high, lies where it slid toward the basin 270

Figure 1.1 A view of the Guadalupe Mountains from the south. The massive cliff is limestone, part of the reef complex that grew here and throughout west Texas during the Permian. The older of the two pulses of extinction ended the deposition of the Permian reef.

million years ago. Then the trail climbs steeply, switching back and forth until the massive limestones of the reef finally appear some three miles and eight hundred feet above the parking lot.[1]

Entombed here is the world of the Permian, the very last profusion of life before the extinction. Water is always precious in west Texas but whenever I climb the Reef Trail I take enough to pour on the rocks. A splash of water and the fossils spring out of the whitish-gray limestone, opening a window into a world of animals far different than we find in modern oceans. Some rocks seem to consist of nothing but half-inch-long grains of rice: the skeletons of single-celled organisms known as foraminifera. Forams were abundant and evolved rapidly, providing a rough clock to the age of these rocks. Brachiopods are common too. With two valves they look superficially like clams, but brachiopods on the half-shell would be hard to swallow. In place of the muscular and tasty foot of a clam, the shell enclosed a looping curtain of filaments that filtered food from the water. Distant cousins to clams and other molluscs, the hundreds of brachiopod species helped build much of the reef. Their cousins the bryozoans share the filamentous filter-feeding structure of brachiopods, but formed tightly packed colonies growing as lacy fronds, massive stony buttresses, and intricate fans (figures 1.2, 1.3).

Today sea urchins, sand dollars, and starfish move easily, and many of them are happy carnivores. But the five living groups of echinoderms (starfish, brittle stars, sea urchins, crinoids, and holothuroidians, or sea cucumbers) provide a restricted view of the evolutionary history of the group. Starfish and sea urchins were uncommon during the Permian, while echinoderms that lived attached to the sea bottom were abundant. A long stack of circular calcite plates formed a stalk that attached crinoids to the sea floor. The body of the animal sat atop the stalk, encased by a network of plates and surrounded by a circle of arms forming an effective net for filtering organic material out of the water.

Brachiopods, bryozoans, and crinoids had dominated the world's oceans for 250 million years, attached to the sea bottom and placidly filtering small animals out of the ocean water. None of them were active predators, could move as adults, or had much

Figure 1.2 A reconstruction of the Permian reef of west Texas based on over forty years of research at the National Museum of Natural History. A school of nautiloids with a straight shell are passing in front of the reef. Most of the animals are various brachiopods, along with some bryozoans, sponges, and clams. Corals, one of the major modern reef builders, were almost absent from this Permian reef. Smithsonian Institution photograph.

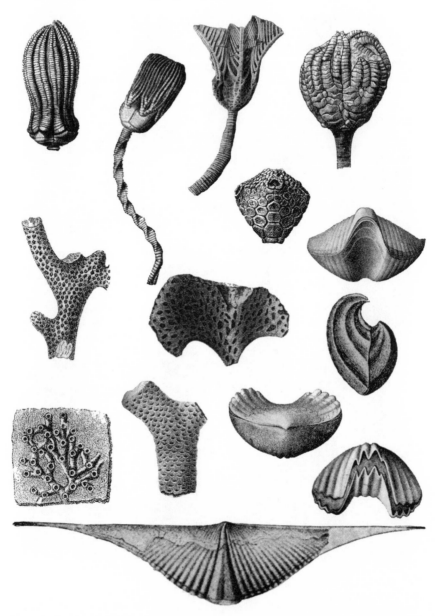

Figure 1.3 The dominant invertebrate marine animals of Paleozoic oceans were crinoids (upper four figures), colonial bryozoans (the stick and lacy fronds on the left) and brachiopods (the four bivalved shells in lower right). Grabau (1904), Wanner (1930, 1931).

meat to them. The *Joy of Permian Cooking* would have been a short book indeed. Their world was about to end.

. . .

This book is about what had happened in the 20 million years separating the rich faunas preserved in West Texas and the scrappy, grubby fossils of the Early Triassic of Utah. Between the reefs of Texas, teeming with fossils and rivaling modern reefs in the number of species, and the barren deposits in Utah lies the greatest biodiversity crisis in the history of life—a far greater crisis than the extinction of dinosaurs 65 million years ago. What so turned the world upside down that biodiversity plunged from hundreds or even thousands of species at a single locality, with thousands of individuals, to perhaps only a dozen species?

Between the rocks in Texas and Utah the Earth experienced two great crises some 10 million years apart. These twin disasters extinguished at least nine of every ten species in the oceans. Moving up the taxonomic hierarchy to more inclusive groups, about 82% of genera and fully half of all marine families disappeared, a level of extinction that dwarfs any of the other great mass extinctions. As many marine families were wiped out during this event as the next two largest mass extinctions combined. Each of these Permian events alone was greater than the extinction that killed off the dinosaurs. Together the twin crises form the greatest biotic catastrophes of the past 543 million years (figure 1.4).

On land, plants and animals came closer to complete elimination than at any point since they first evolved, and yet they rebounded in a few tens of millions of years. These crises at the end of the Permian Period were so extreme, and the animals on either side of the events so different, that John Phillips and other mid-nineteenth century geologists took them as evidence for separate creations of life. Today geologists recognize the Permo-Triassic boundary as a fundamental turning point in the history of life, bringing the world of the Paleozoic to a close, and, in the aftermath of the extinction, constructing the world of today. Despite all the evolution of the past 251 million years, today's oceans still reflect the winners and losers of events at the close of the Permian.

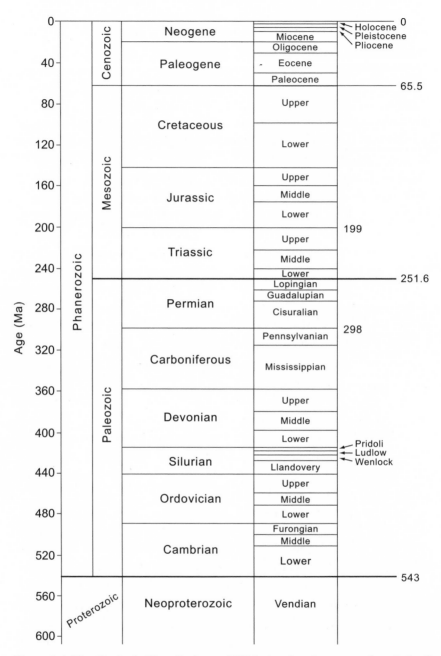

Figure 1.4 The Geologic Time Scale as of 2004, showing the eras and periods of the Phanerozoic (the past 543 million years) and the current ages assigned to the boundaries between the periods.

What could have triggered such a massive loss of life? Geologists are endlessly creative so there is no lack of suggestions, some of which are even plausible. Not wanting to feel left out of the mother of mass extinctions, physicists, evolutionary biologists, and complexity theorists have all offered their solutions to the mystery. And it is a mystery. Over the past several decades, dozens of different hypotheses have been proposed, ranging from the movement of the continents, the warming of the seas, and the impact of an extraterrestrial object to the eruption of massive volcanoes or the suggestion that ecosystems had become too complex and collapsed with no physical trigger at all. Unlike other mass extinctions where the number of physical events is relatively few and causality is easy to establish, so many things occurred during the Permo-Triassic interval that establishing what caused the extinction is very difficult.

The simple truth is that we do not know what caused these twin extinctions. Or at least I don't know. Some of my colleagues are less reticent (or more confident) but there is certainly no agreement yet. None of the extinction models fits all the evidence and some hypotheses require data that despite every effort have not been found. Ten years ago the end-Permian extinction appeared to be a prolonged, drawn-out event over millions of years. Better data on fossil occurrences through the Permian have revealed two discrete episodes of extinction in the sea: one at the end of the Middle Permian, and the second at the very end of the Permian. Although I will touch on the older crisis at several points, particularly in chapters 5 and 8, we know relatively little about it. This first extinction pulse killed a vast number of species in the oceans, probably occurred during a sudden drop in sea level, and coincides with a sharp shift in the cycling of carbon through the oceans and atmosphere. Beyond this we do not even know if plants and animals on land also suffered extinction. This first extinction pulse seems to have had little long-term impact on life, for many Late Permian fossil assemblages look much like those before the extinction.

The greater enigma is the second, catastrophic extinction at the end of the Permian. This shattered Paleozoic ecosystems so thoroughly that they never recovered, and this is the problem to which I have devoted the past twenty years of my research. Unless I specifically mention the earlier extinction when I discuss the Permian extinction, I am referring to this latter episode. Among the myriad suggestions, four stand out as possible causes: the impact of a meteorite or comet; climatic destruction from massive volcanism in Siberia; the oceans losing their oxygen (becoming anoxic) and snuffing out the animals that require it; and a combination of several interacting and mutually reinforcing events.

Almost everything we know is consistent with an extinction caused by the collision of an extraterrestrial object at the very end of the Permian: a very rapid extinction, a dramatic shift in the flow of carbon through the oceans and atmosphere, and extinctions on land and ocean. We will discuss the evidence for the impact of a meteorite triggering the end of the dinosaurs in the next chapter, but in contrast to this end-Cretaceous event, there is only suggestive evidence for impact in the Permian. Permo-Triassic sediments do not contain a blanket of debris that rained out from the dust cloud of an impact. There is no sign of characteristic extraterrestrial elements such as iridium. Some geologists have offered tantalizing suggestions of impact, but so far these have failed to convince most scientists.

The end-Permian mass extinction coincides with one of the most massive volcanic eruptions of the past 600 million years. It is hard to believe this is a coincidence, but just how volcanism triggers a mass extinction is unclear. Global cooling from erupted dust, followed by global warming from clouds of carbon dioxide and acid rain from billowing sulfur are commonly proposed links between volcanism and extinction, but are very difficult to test. A number of scientists, largely physicists but including some geologists who should know better, have proposed that an impact triggered the volcanism. While I admire the broad sweep of such generalizations, there is little geological support for it.

Geologists have uncovered considerable evidence for anoxic waters in both the deep sea and shallow water near the Permo-Triassic boundary, and this has led to the third leading hypothesis: the spread of low-oxygen, or anoxic, waters. What has never been clear is how low-oxygen waters cause widespread extinctions of land plants, insects, and vertebrates. If anoxia was a major cause of extinction, it must have been linked to some other process responsible for extinctions on land.

The final possibility is that this event, like much of history, was messy and lacked a single cause. In 1993 I termed this the *Murder on the Orient Express* hypothesis in honor of the wonderful Agatha Christie mystery where the solution is that *all* the suspects participated in the crime. Some of my scientific colleagues tend to dislike this idea. Not because it may not be true, but because science is about testing ideas and rejecting those that fail the test. Single causes are much easier to test than multiple causes and thus, to some, seem more "scientific."

Ten years ago I was opposed to the first three possibilities because they seemed inconsistent with our evidence for the speed of the end-Permian mass extinction and with the timing of other geological events. Despite much effort since 1980, no evidence for impact had been found at any Permo-Triassic boundary site, and the pattern of extinction appeared to be too prolonged to be due to an impact. The volcanism model bothered me because I could not see a causal link to the extinction. It also appeared that volcanism could not explain the well-documented shift in the carbon cycle associated with the extinction. Then as now, the various anoxia hypotheses (there are at least three) suffer from an inability to explain the terrestrial extinction. There was a difficulty with causality too: was the apparent evidence for anoxia better explained as a consequence of the mass extinction?

I have changed my mind as our research has dramatically shortened the duration of the extinction. I confess my hope that the cause turns out not to be an impact for an aesthetic reason. The Permo-Triassic and Cretaceous-Tertiary mass extinctions look increasingly similar. Since we already (believe we) know that impact can cause mass extinction, we would learn more about life's susceptibility to catastrophe if a very different, essentially earthbound process produced the end-Permian crisis. The physicists, seeking generality, hope for impact while the historians, celebrating diversity, yearn for more intrinsic events.

· · ·

This book is frankly written as a mystery story. Chapter 2 introduces the suite of possible perpetrators of the mass extinction and outlines the type of evidence geologists and paleontologists need to evaluate each hypothesis. Chapters 3 to 7 detail the actual evidence we have to test the hypotheses presented in chapter 2, in essence providing the clues needed to eliminate some suspects. By the end of chapter 7 most readers will probably have their own ideas as to the cause of the extinction. Had I wanted, I could have skewed the material to favor one hypothesis or another (and I expect some of my colleagues will claim I have). As far as possible, I have attempted to put the best light on each of the potential miscreants in chapter 2, and have provided all the relevant clues in chapters 3 to 7. The denouement occurs in chapter 8 where we return to the hypotheses and address the cause of the extinction. I present my views of the cause, but since at the moment it is not clear what caused the extinction, no doubt some readers will arrive at very different conclusions. The final section, chapters 9 and 10, discuss the recovery following the mass extinction and its impact on the history of life.

Science is not simply about developing plausible ideas and telling a good story, but about developing ideas that are congruent with the facts as we know them, then critically testing them by collecting new information. Chapter 2, A Cacophony of Causes, sketches the leading causes for the end-Permian mass extinction and identifies the kind of data required to evaluate them. Since some understanding of the other mass extinctions is useful to place these hypotheses in context, this chapter begins with a brief précis of these other events.

The keys to understanding the end-Permian mass extinction lie in China. There are more marine rocks spanning the Permo-Triassic here than in the rest of the world combined and Chinese geologists have been at the forefront of establishing our new understanding of the Permian. I have been fortunate to collaborate over the past decade with an excellent group of Chinese geologists at the Nanjing Institute of Geology and Palaeontology. Rocks at Meishan, just south of the Yangtze River between Shanghai and Nanjing, are the global reference point for the Permo-Triassic, and an account of our fieldwork there and elsewhere in China forms the basis of South China Interlude, chapter 3. This chapter also introduces the complexity of the fossil record. One of the most useful pieces of geological evidence at the Permo-Triassic boundary is the abrupt shift in the carbon cycle. This is such a useful tool that I introduce it in chapter 3, employ it to correlate to other regions in chapter 4, and then discuss the causes and interpretation of the signal in chapter 7.

Evaluating the various proposed causes of the extinction depends upon knowing how rapidly the mass extinction occurred: no dates, no rates, as a colleague of mine is fond of saying. A mass extinction drawn out over millions of years demands a much different explanation than an extinction lasting only a few hundred thousand years, or less. Establishing the duration of the extinction is critical. Together with my good friends Jin Yugan and his colleagues from Nanjing, and Sam Bowring and his group from Massachusetts Institute of Technology (MIT), we have scoured south China in search of volcanic ash beds bracketing the extinction interval. Chapter 4, It's a Matter of Time, introduces the tools used to date volcanic eruptions. We have shown that the end-Permian destruction occurred in less than 500,000 years and probably less than 160,000 years. We cannot yet say how much less than this, but nothing excludes the possibility that the extinction was as abrupt as the end-Cretaceous mass extinction. But this only tells us how fast the extinction was in South China. This chapter concludes with a look at how we can extend these results to other regions by integrating them into correlation schemes based on fossils and the distinctive shifts in the carbon cycle.

Chapter 5, Filter Feeding Fails, chronicles the shift in perspective over the past decade from a lengthy event to the twin crises, and the patterns of extinction and survival in the oceans. Not all groups suffered to the same extent, and these differences provide important clues to the causes of the extinction. The varying patterns of extinction and survival between different groups are another important clue in determining the causes of the extinction.

South African Eden, chapter 6, focuses on the extraordinary fossil animals of the Karoo Desert of South Africa, one of my favorite places. Here a magnificent record of vertebrates preserves documentation of extinctions near the Permo-Triassic boundary. This chapter also discusses extinctions among plants and other animals, providing critical clues to whether life on land was as severely affected as life in the oceans. Absent such information we cannot distinguish between proposed causes that only affect the oceans, and those with more global impacts. Insects are remarkably resistant to extinction, and as far as we know, the only mass extinction insects ever suffered occurred during the Permo-Triassic. Other major disappearances occurred among land vertebrates and some plants, although it is not entirely clear whether the events on land and in the oceans occurred at the same time.

Life depends on the flow of carbon through the oceans and atmosphere, and its burial as organic debris, oil, and limestone. Consequently, geochemists have developed very sensitive tools to follow perturbations to this cycle. The problem is that there are often too many ways to generate these shifts, as discussed in chapter 7, The Perils of Permian Seas. We know that a massive volume of carbon from organic matter was added to the atmosphere and oceans at the Permo-Triassic boundary. Where did the carbon come from? The simplest solution is that it came from all the organisms that died during the extinction, but it turns out that if every living thing were vaporized today the resulting carbon is not enough to cause the shift seen at the Permo-Triassic boundary. So some of the carbon must have come from some other source, such as coal beds, methane trapped in deep-sea sediments, or carbon dioxide from volcanoes. In addition, there appears to have been a longer-term shift in the amount of carbon buried in rocks from the Permian into the Triassic. The elements sulfur, oxygen, and strontium each have a cycle that was perturbed during the Permo-Triassic transition. In chapter 8 we return to the various hypotheses discussed in chapter 2 and evaluate them.

· · ·

The aftermath of the extinction may be as puzzling as the mass extinctions, and from the standpoint of the history of life, they are at least as important. After the other mass extinctions, new species begin to proliferate within a few hundred thousand years. Chapter 9, Resurrection and Recovery, charts the cryptic patterns and processes of the end-Permian aftermath. Despite the nine species of gastropod I could collect beneath the brush and scrubby piñon pine of that Utah hillside, I know of two dozen snail lineages that survived the end-Permian extinction, but that simply cannot be found anywhere in the world during the first 5 million years of the Triassic. These lineages clearly survived the extinction, but are missing until they reappear, phoenixlike, from the ashes of the crisis. Such seemingly miraculous rebirths were christened Lazarus taxa by David Jablonski, a paleontologist from the University of Chicago (and normally not the most religious of men). The number of Lazarus lineages among gastropods and other taxa tell us that many fossils simply were not preserved in the fossil record. But something else was also happening. Lazarus taxa do not reappear in the fossil record until the pace of diversification, or origination of new species, begins to pick up. This suggests some connection between the recovery and the reemergence of the Lazarus taxa. What? There are two explanations: The first is that environmental perturbations continued for millions of years after the extinction, retarding the recovery. The alternative is that the extinction so disrupted the ecology of communities that a slow convalescence was required before ecological communities could begin to function, new species appear, and Lazarus taxa emerge.

To grasp the significance of the extinction for the history of life, we need to understand why those hillsides in Utah have so few fossils, and why rocks all over the world have the same grubby fossils. Out of such scrappy remains and the return of the Lazarus taxa, evolution molded the world of today. Many environmentally minded people view extinction as tragic, even intolerable, an assault upon the integrity of the ecosystems that sustain us, and likely to produce a world we would not want to pass on to our children's children. At the rapidity of our current, human-induced biodiversity crisis, extinction is intolerable. But as discussed in the final chapter, The Paradox of the Permo-Triassic, mass extinction is a powerful creative force. From the wreckage of mass extinctions the survivors are freed for bursts of evolutionary creativity, changing the dominant members of ecological communities and enabling life to move off in new and unexpected directions.

Before anyone rushes out to proclaim that mass extinction is a good idea, remember how long the recovery lasted. The 5 million years after the end-Permian mass extinction were a pretty lousy time to be alive (even for snails), and the diversity of life in the seas did not begin to approach preextinction levels for tens of millions of years. A long time to wait.

CHAPTER 2
A Cacophony of Causes

As Europeans began exploring the world in the seventeenth century they came upon the ruins of great civilizations: the pyramids and Great Sphinx in Egypt, Ur in Mesopotamia, and eventually the New World Mayan ruins of Copan and Tikal, and Machu Pichu of the Inka. These complex civilizations and other vanished centers like Mesa Verde of the *Ancient Puebloans* in the American Southwest or the Moundbuilders of the Mississippi Valley continue to fascinate archeologists and the general public. Something deep within us (fear, perhaps?) renders stories of decline far more attractive than the long, preceding periods of success. Explanations for the collapse of such societies run the gamut from warfare and political intrigue to plagues, environmental degradation and other disasters. Despite the rich historical and archeological records of these societies, discriminating between competing hypotheses is frustrating. Some archeologists have championed a single cause upending a society, while others have adopted a more nuanced approach, constructing intricate webs of interacting causes.

Similarly, paleontologists look at mass extinctions as extraordinary events demanding amazing and wondrous causes. But as with more recent human events, there are often so many competing geological, climatic, and chemical events that establishing which represents the cause of the extinction and which the effect is a real challenge. This chapter considers the myriad different causes that have been proposed for the Permian-Triassic mass extinction, from volcanism to climate change and low oxygen in ocean waters. Ideally, from each hypothesis should flow testable predictions. If an extraterrestrial impact occurred, for example, the resulting mass extinction must have been very rapid, nearly simultaneous around the world, and we should be able to find some evidence of the impact (unless it was one of those stealth impacts of which a few commentators are so enamored). Some of my favorite hypotheses, so aesthetically pleasing that they really ought to be true, are either clearly wrong or cannot even be tested. In understanding the reasons for the collapse of ancient civilizations a comparative approach is useful for providing a sense of the range of plausible explanations. If there is no evidence that hurricanes destroyed any other tropical civilization, they are an unlikely explanation for the fall of the Maya. So before setting off through a cacophony of competing hypotheses for the Permian mass extinction, I begin with a quick tour through other biodiversity crises and their causes.

· · ·

The Permian was "discovered" by Sir Roderick Impy Murchison in 1841 in the rocks of the Ural Mountains of Russia. Far from impy, Murchison was the imperious director of the British Geological Survey and a powerful presence in mid-nineteenth-century European geology. British colonialism knew no bounds, and Murchison was keenly interested in proselytizing for the "right" sort of geology where one of the principal goals was delineating rocks encompassing units of geologic time. He had previously recognized other time units, including the Silurian, a distinctive suite of rocks in the Welsh borderlands that he named after the ancient tribe that once inhabited the region. Gradually these distinctive packages of rock and their characteristic fossils were assigned names, often after the ancient tribes where the interval was first recognized. Murchison visited Russia in the 1840s at the invitation of the czar and identified a sequence of rocks and fossils in the Ural Mountains that were clearly different from the older Carboniferous rocks of Britain, or the younger Triassic rocks of Europe. These became the Permian, named after an ancient kingdom in the area that also gave its name to the city of Perm. (I wonder how many Texans knew, during the height of the Cold War, that Permian High School in Midland traced its name to the Russian heartland.) The geologist von Alberti had described a series of rocks in the Alps as the Triassic, and it soon became evident that these lay just above Murchison's Permian. Through this process of identifying packages of rocks with distinctive fossils, by the late 1800s European geologists established the relative order of different geologic units into the geologic timescale. Today we know the Permian stretches from about 298 million years ago to 251.6 million years ago, but neither Murchison nor von Alberti had any sense of this.

Murchison and von Alberti's correlations between Russia, the Alps, and England, indeed all the correlations made by geologists across Europe during the middle 1800s, relied on fossils. Following a century or so of confusion geologists had recognized that similar rocks were produced in similar environments, sandstones by rivers and shorelines, shales in offshore basins, carbonates near reefs, and so on. There was no reason to believe that two sandstones, even two red sandstones, were of the same age. With the realization that particular kinds of fossils occurred only during a narrow span of time came the discovery of correlation: the use of these fossils as a basis to infer that the rocks in which they were found were formed at the same time. From this was born the ability to map geologic units in space, to order rocks in time, and ultimately to develop a sense of the history of the earth. If fossil A is always found beneath fossil B in England, and B always lies beneath fossil C in France, then the rocks containing A must be older than those containing C. Through a series of such hypotheses, tested and retested against reports from the field, the geological time scale was established.

The geologic time scale was largely based on the rocks and fossils uncovered in Britain and Europe, but because the extinctions reflect global events it would look much the same had it developed in China, Canada, or Peru. There are a few exceptions. Nothing much happened during the Silurian and it differs little from the Lower Devonian, so it is hard to consider the creation of the Silurian Period as anything but an unfortunate mistake. Similarly, the division between the Carboniferous and Permian is so obscure in the United States that American geologists argued over the dividing line into the 1960s. Nonetheless, the fossils of the Cambrian to Permian share far more in common than they do with those of the Triassic through Cretaceous, or with the post-Cretaceous.

John Phillips was the first geologist to recognize the implications of this pattern. In 1840 Phillips divided the fossil record into the Paleozoic, Mesozoic, and Cenozoic eras. The fossils of each era were so distinct—the brachiopods, crinoids, and bryozoans of the Paleozoic; some bivalves of the Mesozoic; and the gastropods, crabs, and other bivalves of the Cenozoic—that Phillips was convinced each represented a separate creation of life. This was a perfectly logical assumption in Phillips's world. Letters between Phillips and Darwin make it clear that he never adjusted to Darwin's theory of evolution. In 1860 he drew the first illustration of the diversity of life through time (figure 2.1). The boundaries between the eras, and between many periods, are defined by major turnovers in the dominant fossils. In other words, by mass extinctions. Phillips first considered the relative durations of the different periods, but along with other geologists and physicists in the mid-1800s, he was soon caught up with trying to estimate how much time in years (what geologists call absolute time) was occupied by different periods. He focused on the amount of time required to deposit and erode sediments, and his 1860 estimates of the relative durations of the three eras are remarkably close to our present understanding based on precise radiometric dating. In fact, Phillips's estimate of the absolute age in years of the Cretaceous was far more accurate than Darwin's.

Animal fossils only extend through the last 550 million years of the earth's 4.6 billion year history. During this span, paleontologists have identified six great mass extinctions[1] (figure 2.2), although two of these events may be more artificial than real.

We will begin by working backwards in time from the best-known mass extinction, the end-Cretaceous event 65 million years

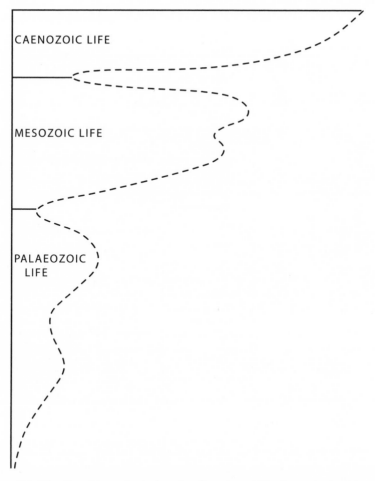

Figure 2.1 The first figure depicting the diversity of life through the Phanerozoic, published by John Phillips in 1860. Note the major drops in diversity corresponding to the end-Permian and end-Cretaceous mass extinctions. Phillips did not concern himself with the extinctions as such, and it would be almost a century before paleontologists began to explore their causes.

ago when the environmental effects of the impact of a large, extraterrestrial object in the Yucatan Peninsula of Mexico wiped out the dinosaurs. In the Late Cretaceous dinosaurs dominated the land, and after some 150 million years of evolution had diversified widely. Mammals were widespread, but the great diversification of modern placental mammals would not occur until after the extinction. Many different kinds of flowering plants had appeared

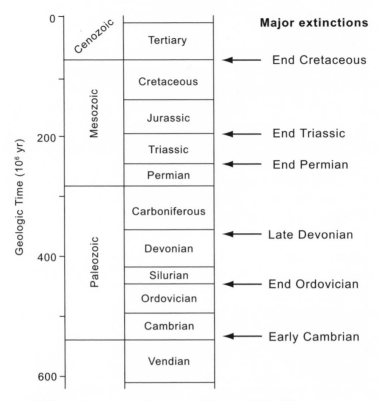

Figure 2.2 The major mass extinctions of the past 600 million years.

in only a few tens of millions of years of the Late Cretaceous. Ocean life bore many relationships to modern ecosystems, with the critical exceptions that corals were relatively insignificant and Late Cretaceous reefs were built by massive, conical bivalves known as rudists, and the coiled ammonite cephalopods flourished as major predators. The extinction eliminated some 47% of marine genera and 16% of families, including several microfossil groups, some bivalves, and the ammonites. Mososaurs and other marine reptiles became extinct too. Plants suffered considerable turnover, with one analysis from the northern Rocky Mountains suggesting a 79% extinction. So much more is known about this extinction than any of the other crises that it is worth considering it in greater detail.

The suggestion that a massive asteroid hitting the Yucatan Peninsula was the trigger has revolutionized geology. For over a century geologists had banished any discussion of catastrophist or extraterrestrial influences on geology, and particularly on the history of life. When in 1980, the father and son pair of Louis and Walter Alvarez and their colleagues from the University of California, Berkeley, linked the end-Cretaceous mass extinction to an extraterrestrial impact, they were met with skepticism and even scorn by most paleontologists, and for fairly good reasons. Geological training since the mid-1800s had emphasized the primacy of slow, gradual, processes and rejected catastrophic explanations, particularly extraterrestrial ones. Geologists were not blind to the existence of meteorite impacts—many had been identified during the geological fieldwork spurred by the Apollo program—and catastrophist theories were proposed from time to time. But speculations connecting impacts and major events in the history of life were either ignored or discounted. Geologists knew (or thought they knew) that was not the way the world worked. Although some physicists viewed this as simple obstinacy, in 1980 the pattern of extinction appeared gradual. Paleontologists reasonably expected that a catastrophic mass extinction should produce an abrupt and catastrophic disappearance of species in the fossil record.

The critical feature of the Alvarez impact hypothesis was the discovery of a thirty-fold increase in concentration of the element iridium at the Cretaceous-Tertiary boundary at Gubbio, Italy, and at other sites. Earlier work had suggested that iridium in sedimentary rocks came from the constant rain of micrometeorites. Louis Alvarez suggested to his son Walter, a professor of geology at Berkeley, that the changing abundance of iridium in marine sediments might be a useful index for changes in sedimentation rates. Assuming the influx of extraterrestrial debris was constant, if the rate of sediment deposition slowed, then the abundance of iridium in a given thickness of sediment should increase (since that layer of rock would be sampling the extraterrestrial debris deposited over a longer span of time). When Walter Alvarez and his colleagues applied this to the Cretaceous-Tertiary section at Gubbio, they found an incredible spike in iridium concentration right at the boundary. Moreover, they uncovered a similar pattern at two other sites, in Denmark and New Zealand, indicating that whatever caused the iridium spike, the pattern seemed to be global. They realized that the most plausible explanation was not a change in the rate at which the sediments had been preserved; the increase in iridium was too great. Instead they proposed that the impact of a meteorite or comet ended the reign of the dinosaurs.[2]

Whether it was a meteorite or a comet, the object that struck the Yucatan Peninsula was some 10–15 kilometers in diameter. Since 1980, field and laboratory studies and computer simulations have revealed much of what happened 65 million years ago. The impact excavated a cavity about 100 kilometers in diameter, displacing some 100,000 cubic kilometers of debris, including the melted rock from the immediate impact, vapor, and solid debris. As the first cavity collapsed, a crater 170 to 300 kilometers in diameter formed (since it is deeply buried under younger sediments, it is difficult to be sure of the size). Beyond the massive earthquakes, the impact triggered tsunamis that sped out from the Yucatan. The blast effects incinerated everything within thousands of miles and extended into the western United States. Rocks blasted out by the impact heated on reentry through the atmosphere and triggered wildfires on many continents. But the most far-ranging effects were from the rocks vaporized by the impact. In North America the impact deposit from the Cretaceous-Tertiary event can be readily separated into two layers: a lower layer of material directly ejected from the impact and an upper layer of atmospheric fallout from material in the vapor-rich plume that rose far into the atmosphere and was then carried around the world. Only the upper, fallout layer is found in Europe, Africa, and the southern hemisphere, although it thins away from the impact site.

The dust of the vapor cloud obscured the sun and cooled the earth for months, probably shutting off photosynthesis for a lengthy period, and causing much of the extinction. Sulfur from the carbonate rocks in Yucatan would have been turned into sulfuric acid and then acid rain. How long this destruction lasted is unclear. Early workers suggested the dust cloud persisted for years, but later computer simulations showed that as the particles began to stick to each other they would rain out of the atmosphere more rapidly.

The site of the impact was unknown in 1980 and some geologists doubted it would ever be found. On the Moon and Mars the lack of wind, rain, and active tectonics preserves craters for billions of years. Here on Earth the trace of even an immense impact are soon obliterated by erosion and buried by sediments. The surface of the Yucatan Peninsula in Mexico betrays few signs of the impact. But impacts do leave a footprint in gravity and magnetic patterns. The Earth's gravity and magnetic field vary across the surface of the earth. Mountain ranges have deep roots of low-density rock that project down into the higher-density mantle. This allows mountains to "float" on the mantle. This difference in mass between a mountain and the overall global average can be detected by aerial surveys of the gravity and magnetic fields, with the mountains showing up as a negative gravity anomaly: since the deep root of the mountains displaces higher-density mantle rocks. As a mountain erodes, the deep root rises to compensate, and the negative gravity anomaly still shows up. An impact alters the regional gravity and magnetic patterns much like a mountain range, and old, eroded craters can be found in the same way. Gravity surveys revealed an unusual circular structure in the Yucatan but only later was it identified as a possible impact structure. Today Chixulub had been widely accepted by geologists as the site of the Cretaceous-Tertiary boundary impact, although some controversy remains over whether it exactly corresponds with the mass extinction horizon.

Iridium was soon picked up at many other Cretaceous-Tertiary marine and terrestrial boundary sections around the world. But the analysis of iridium turned out to be far more difficult than initially expected. This was not terribly surprising, since geologists were looking for vanishingly small amounts of iridium, often at the absolute limits of what could be detected. In trying to detect such minuscule amounts, random fluctuations, small differences in analytical procedures, and inadvertent contamination can be the difference between finding iridium and coming up empty. Having a second, independent indicator of an extraterrestrial impact would obviously be ideal, particularly since some extraterrestrial objects, such as comets, are unlikely to have much iridium. Geologists soon established another sign of impact: a characteristic pattern of parallel fracture sets within mineral grains, particularly quartz. Commonly known as shocked quartz, these fractures are caused by the tremendous energy of impact and the associated pressure wave. Just as the unambiguous detection of iridium required careful laboratory work, reliably identifying shocked quartz also turned out to be more difficult than expected, and there have been many false reports of shocked quartz.

I have covered the effects the Cretaceous-Tertiary impact in so much detail for two reasons. First, we know more about this extinction and its likely causes than any of the other events, and second, because impact is also a possibility for the end-Permian event. The events of the end-Cretaceous provide useful comparison for what we might expect in the Permian.

The end-Triassic mass extinction 199 million years ago has been well studied in Europe, but the extent of the crisis in other parts of the world has been unclear. A rapid rise in sea level coincides with the elimination of some 53% of genera and 22% of marine families. Particularly affected were cephalopods (with octopus and pearly nautilus as modern representatives), brachiopods, clams, and some snails. Many of the Paleozoic groups that survived the end-Permian mass extinction finally disappeared. About 12% of vertebrate families also disappeared, and in the aftermath, dinosaurs took command of the land. Some paleontologists have championed the spread of anoxic waters as a cause and others have claimed evidence for an extraterrestrial impact but recent high-precision radiometric age dating of volcanic rocks along the east coast of North America, in Morocco, Spain, and Brazil has shown that the mass extinction coincided with a series of massive volcanic eruptions. These areas were then united in the supercontinent of Pangaea, and the volcanism may be associated with the opening of the Atlantic. The leading candidate for the trigger of the end-Triassic extinctions is this massive volcanism, known to geologists as the Central Atlantic Magmatic Province because it spans the Atlantic margin.

The next mass extinction in the series is a complex and still poorly understood episode 376 million years ago during the Late Devonian. Some 57% of marine genera and 22% of marine families disappeared during a series of events. Again, brachiopods experienced considerable extinction as did trilobites, corals, and other members of reef communities; bivalves and gastropods suffered relatively minor losses. In Europe, some paleontologists see evidence for at least three discrete extinction pulses; global studies are more equivocal. At least one extraterrestrial impact is known from the Late Devonian, but it does not correspond to any extinction horizons. Possible causes include spread of low oxygen waters, changes in the chemistry of the oceans, and climate change. But until geologists learn more about the extinction, it may be premature to consider the causes.

The second largest mass extinction brought the Ordovician Period to a close 439 million years ago. The trilobite-dominated communities of the Cambrian disappeared during the Ordovician Radiation, diluted by the rapid growth of the dominant Paleozoic communities with articulate brachiopods, crinoids, bryozoans, and some Paleozoic corals. These groups lived attached to the sea floor and filtered food out of the water (there was little, if any, life on land this early in animal history). Two discrete extinction events separated by perhaps 500,000 to 1 million years extinguished 60% of marine genera and 26% of marine families. Glaciation and global cooling were the most likely causes of this event. The extinction began with a rapid glaciation and consequent drop in sea level, and then as the glaciers melted, sea level rose rapidly and delivered anoxic, or low oxygen, waters into shallow seas and caused the second pulse of extinction. Brachiopods suffered particularly heavy extinction, as did trilobites, echinoderms and bivalves, and some corals. One of the most enduring curiosities of this episode is why it had such little long-term ecological impact. The various surviving groups produced many new species over the following 1–2 million years, and Early Silurian communities do not appear to be much different from those before the extinction.

Stephen Jay Gould made the great Cambrian explosion of animal life famous in his book *Wonderful Life*, describing the "weird wonders" of the Burgess Shale. The Cambrian Radiation marked the origins of diverse animal fossils, from the ever-popular trilobites through many early shelled organisms. The oldest recognized mass extinction in the fossil record occurred shortly after this, about 512 million years ago, near the end of the Early Cambrian. This event has only been recognized during the past decade and appears to have eliminated about 50% of all marine species, including many of the distinctive animals of the Early Cambrian seas. An odd, spongelike group called the archaeocyathids disappeared, as did some trilobites and a variety of tiny shelly fossils. The Early Cambrian was a time of rapid overturn in biodiversity, with high rates of extinction and origination. The disappearance of so many distinctive lineages at the close of the Early Cambrian suggests that something different happened at this point. Unfortunately we know so little about this event that we can say little about the likely causes.

This brief survey shows that there are many different ways of causing mass extinction: global cooling in the Late Ordovician, massive volcanism across the Triassic-Jurassic boundary, and the impact of an extraterrestrial object in the Late Cretaceous. Finally, the causes of two events, in the Late Devonian and Early Cambrian are unknown, and the series of Late Devonian events are much different from other episodes. At least among the great mass extinctions no evidence of a single, unified cause has been uncovered, although as we will see, suggestions persist of links among less drastic biodiversity crises.

· · ·

While the relatively abrupt changes in fossil were evident to John Phillips and his colleagues in the mid-1800s, these transitions appear to have been of little interest to geologists and paleontologists until the 1950s. I am still amazed that Phillips and several subsequent generations of geologists were so uninterested, but it simply never occurred to them to inquire into the events behind the extinctions.

Not all major boundaries in the geologic time scale are demarcated by mass extinctions. The suggestion that a mass extinction occurred at the end of the Triassic but not at the end of the Jurassic was primarily the work of the late Jack Sepkoski of the University of Chicago. While in graduate school at Harvard, Jack began the Herculean task of assembling information on the oldest and

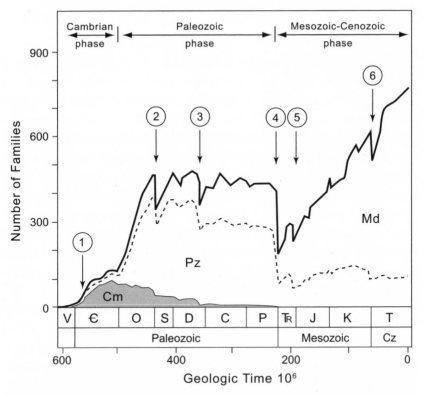

Figure 2.3 Marine diversity through the past 600 million years, modified from a figure by Jack Sepkoski (1984), based on his compendium of the first and last occurrences of marine families. The shaded area denotes the number of families assigned to the trilobite-dominated Cambrian Evolutionary Fauna. The dotted line shows families comprising the brachiopod-bryozoan-crinoid dominated Paleozoic Evolutionary Fauna. This fauna suffers particularly harsh extinction at the close of the Permian, and the bold black line denotes the number of families in the Modern Evolutionary Fauna, including snails, clams, crabs, and sea urchins. The circled numbers are the six great Phanerozoic mass extinctions: 1, Early Cambrian; 2, end-Ordovician; 3, Late Devonian; 4, end-Permian; 5, end-Triassic; 6, end-Cretaceous. The letters correspond to the geologic periods in figure 1.4. Redrawn from Sepkoski 1984.

youngest occurrence of all genera and families ever found as marine fossils. Some broad compilations had been attempted before, but in nothing like the detail Jack devoted to the task (figure 2.3).[3]

In 2001 Richard Bambach and Andy Knoll of Harvard, both of whom will figure prominently later in this account, took another look at the data compiled by Sepkoski. They suggested that the pattern of mass extinctions was more complex. Although the Late Devonian and end-Triassic episodes showed up prominently in earlier studies, they remained enigmatic. Their diversity patterns are distinct: although their extinction levels are high, they are not as exceptional as for the end-Ordovician, end-Permian, and end-Cretaceous. The principal characteristic of the Late Devonian and end-Triassic episodes is a failure of origination rather than exceptional extinction, leading Bambach and Knoll to describe them as "mass depletion" episodes. However this new wrinkle in the story develops, the evolutionary impact of these events is certain. Well-developed reefs vanished for almost 100 million years after the Late Devonian, and the end-Triassic saw off some soft-bodied chordates (animals with a spinal cord, but in this case lacking bones), known as conodonts that are among the key index fossils of the Paleozoic and Triassic.[4]

But are mass extinctions truly different, or is there a continuum of extinction magnitudes between the end-Permian, the largest of the past 540 million years, and lesser episodes? Some extinction has occurred during every interval of geologic time. Dave Raup produced a fascinating chart arraying the magnitude of extinction (figure 2.4). This shows a continuum, downplaying any sharp difference between the mass extinctions and other times. Raup and Sepkoski identified another interesting extinction pattern in 1984 with a very different implication. Again using Sepkoski's magnificent compendium of fossil families, they described a cyclical pattern of about 26 million years to mass extinctions between the end-Permian and the present (figure 2.5). Some of the extinction events were very small, barely above the noise in the data, but the cycle encompassed the end-Triassic and end-Cretaceous mass extinctions, as well as several smaller but previously recognized biotic crises. We will return to this pattern in the final chapter, for it suggests there must be some underlying cause connecting these events, and, moreover, that at least some of these biotic crises may differ from the everyday extinctions in other intervals.[5]

The remainder of this chapter considers the various proposed causes for the end-Permian event. The possible villains, if you will. The scent of an impact is foremost in many people's minds, so

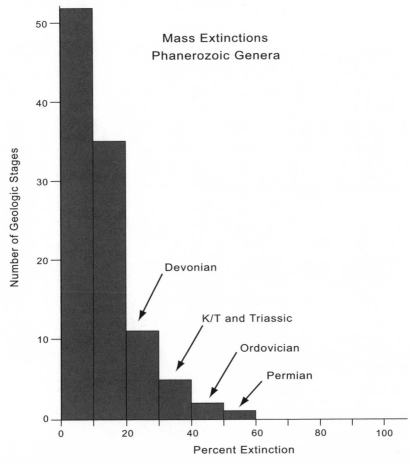

Figure 2.4 David Raup's depiction of extinction intensity, arraying the extinctions in each of the stages of the Phanerozoic by the magnitude of extinction. This shows no distinct differences between mass extinctions and the extent of extinction in other stages. After Raup (1991).

we begin there before turning to other possible extraterrestrial events, and then to the massive flood basalts in Siberia and other types of volcanism and their possible association with climatic changes. This leads us to the role of changing position of the continents and their effect on biodiversity and how the rise and fall of sea level can modify the amount of available area for organisms to inhabit. We turn next to evidence from the carbon cycle for reduced oxygen in the oceans and the possible release of vast amounts of methane. The chapter concludes by describing the

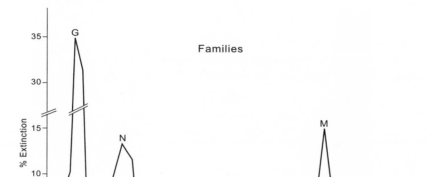

Figure 2.5 Apparent periodicity in extinctions from 250 million years ago to the present, based on peaks in the percentage extinction in Sepkoski's compendium of marine family diversity. Redrawn from Raup and Sepkoski, 1986, used with permission. Copyright 1986 AAAS.

specific predictions from each theory and the kind of evidence we must seek to evaluate them.

• • •

Following the apparent success of the Alvarez hypothesis in ex-plaining the end-Cretaceous mass extinction, the Permo-Triassic boundary was the obvious next place to look for iridium. Volcanic clays at the Meishan boundary sections in south China and another section at Wachapo Mountain, in Guizhou Province, revealed less than 0.5 parts per billion of iridium. Another study by a group of Chinese geologists based in Beijing reported up to 8 parts per billion of iridium from Meishan. Since the Alvarez group had re-ported abundances of about 8 parts per billion for the Cretaceous-Tertiary boundary, the work by the Beijing group appeared to sug-gest that a similar impact occurred at the Permo-Triassic boundary. Yet many attempts to repeat the results of the Beijing scientists proved fruitless. Neither the late Carl Orth at the Los Alamos National Laboratory in New Mexico, nor the group at UC Berkeley could replicate the Beijing claims. Orth's group reported levels of 0.002 and 0.024 parts per billion from splits of the same samples the Chinese group had analyzed. Such vanishingly small abundances are essentially insignificant and provide no evidence for impact.[6]

No one has worked out why the Chinese and U.S. labs got such different results. I asked Carl Orth about this shortly before he died and he suggested analytical differences between the two labs, or even very slight differences in the sedimentation rate between different places. Remember that Walter Alvarez originally undertook his search for iridium to look for differences in sedimentation rates.

Carl Orth and his colleagues did discover two very small peaks in iridium abundance in a core drilled through the Permo-Triassic boundary in the western Alps on the Gartnerkofel. The lower peak coincides with a sharp perturbation in the carbon cycle, and the second occurs higher in the section. The mass extinction is probably coincident with the lower shift in carbon. Pyrite (fool's gold) and other minerals occur with the upper iridium peak, evidence of an unusual chemical environment. Although an impact origin for the iridium could not be ruled out, the overall chemistry of the rocks strongly suggested that the iridium was concentrated by something other than an extraterrestrial impact. More recently a new group of impact experts led by Christian Koeberl in Vienna analyzed these rocks and again found no evidence supporting an extraterrestrial impact. Rather, the slightly elevated concentration of iridium and other elements occurs in a zone of anoxic deposition, which is known to concentrate the elements.[7]

The best proof of an impact would be a crater of the right age. In 2000 several geologists announced the presence of a possible impact structure in the Carnarvon Basin of western Australia, using gravity and magnetic data similar to that uncovered the Chixulub crater in the Yucatan. Australian ranches, or stations, make those in Texas seem puny, and this impact site was named the Woodleigh structure after the station where it was found. To my untrained eye, the gravity data look like a small bull's-eye with a central peak and two, or, if you sort of blur your eyes, maybe three, circular rings. The outermost ring is about fifty kilometers in diameter. High-speed photographs of raindrops falling on water show a rebounding peak from the center of the impact, surrounded by spreading rings. Most of the water in this thin conical peak is not from the drop itself, but from the underlying water. With a sufficiently large impact, the same thing happens on the earth. This central peak represents the resurgence of rock toward the center of impact. But on earth the central uplift can be frozen in place (although never with as sharp a peak as a raindrop). Just such a central peak appears in the gravity map of the Woodleigh structure. Arthur Mory of the Western Australian Geological Survey in Perth and his colleagues argued that the structure is an impact crater at least 120 kilometers in diameter, making it the fourth-largest impact crater known, although still a fair bit smaller than the structure in Yucatan.[8]

Geologists often have great difficulty determining the age of impact structures. Quite fortuitously, a 189-meter-deep well was drilled nearby and produced shocked quartz fragments characteristic of an extraterrestrial impact. Early reports of sediments from an Early Jurassic lake covering the crater, shales with Early Permian pollen in the crater, suggested the crater must be younger than Early Permian, but older than Early Jurassic. Mory and his colleagues provided other evidence the rocks were heated sometime between 280 and 250 million years ago, and suggested this was the age of the impact. So while the impact *could* have occurred at the Permo-Triassic boundary, Mory had no real evidence to connect it with the mass extinction. (This rather significant point was ignored by the press coverage, of course, showing that some science journalists are as susceptible to "spin" as those covering Washington.)

The Case of the Woodleigh Conundrum was resolved in when Wolf Reimold of the University of Witwatersrand in Johannesburg, and Chris Koeberl of the University of Vienna published a critical response to the Mory paper. They questioned the shocked quartz, the estimated size of the structure, and even whether it was produced by an impact. Both Reimold and Koeberl are respected impact geologists, so their concerns were influential. In response Mory and colleagues provided additional data on the possible shocked quartz and convincing evidence for an impact origin for the structure. More important, however, were new radiometric ages pointing toward an impact near the Devonian/Carboniferous boundary, 100 million years before the end-Permian mass extinction. While the Woodleigh structure may warrant further study as an impact structure, there is little reason to connect it with the Permian mass extinctions.[9] Even raising the crater might seem pointless as newer research has revised the earlier hypothesis of a Permo-Triassic date, but this is a cautionary tale to reserve judgment on putative impact features until they are well dated and their impact origin confirmed.

The need for caution continues as I write this book. Early in 2004 Luann Becker of the University of California at Santa Barbara proposed another possible Permo-Triassic impact structure, this time off western Australia where an unusual feature on the sea bottom, the Bedout structure, has garnered attention from geologists. Gravity data, and what have been interpreted as impact debris and impact melts, provide much of the evidence and a single radiometric date of 250.1 +/− 4.5 million years supported assignment of the feature to the end-Permian mass extinction.[10] The Chixulub impact at the Cretaceous/Tertiary boundary demonstrated that a large impact ejects vast amounts of rock into the atmosphere and may trigger massive tsunamis. Both the blanket of airborne ejecta and the jumbled debris of a tsunami are good evidence of impact. The Bedout structure itself, buried deep beneath younger sediment, is an uplifted region of rock evident on seismic imaging of the region. Becker and colleagues interpreted this as the central peak, similar to that in the Woodleigh structure, with the total diameter of the crater similar to Chixulub. It is safe to say that virtually every aspect of this report has been criticized by other impact geologists, from the evidence of impact debris to the quality of the radiometric dates. These doubts have raised strong questions about the significance of the Bedout structure to the Permo-Triassic story, a point to which we will return in chapter 8.

Curiously, the only possible shocked quartz ever reported from the Permo-Triassic boundary comes from eastern Australia, New Zealand, and Antarctica where Greg Retallack of the University of Oregon has been working. Greg is a brash Australian whose modus operandi is often to advocate the opposite of what everyone else believes (a salutary characteristic in science). I was amused in 1996 when he told me that he had found the "magic layer" with evidence of an extraterrestrial impact at the Permo-Triassic boundary. Yet no impact debris has been found in sections in Antarctica and in southeastern Australia; the levels of iridium were insignificant. Evidence for shocked quartz was also ambiguous, since Retallack's quartz grains were smaller and far less numerous than those found at most Cretaceous-Tertiary boundary localities. Although some Cretaceous-Tertiary boundary sites have low iridium levels and few shocked quartz grains, in the absence of any other evidence for impact at the Permo-Triassic boundary, the Oregon group eventually concluded: "Unlike the Cretaceous-Tertiary boundary with abundant evidence of a major impact in Yucatan and globally broadcast ejecta . . . the Permian-Triassic boundary yields only the scent of an impact. Yet, the much more severe extinction at the Permian-Triassic boundary demands evidence of a much larger impact if that were its primary cause. The magnitude and location of impacts at the Permian-Triassic boundary remain uncertain, so their role in Permian-Triassic extinctions remains to be demonstrated."[11]

But impact aficionados should not despair! In 2001 possible new evidence of impact at the Permo-Triassic boundary came from an unexpected source. Readers who were conscious during the 1970s may remember the propagation of geodesic domes across the landscape. Advocated by Buckminster Fuller, these hemispherical structures of intersecting metal rods were the tepees of the Age of Aquarius. In 1985 a group of chemists announced the discovery of a new class of carbon compounds—spheres of sixty or more carbon atoms linked in a lattice of hexagonal and pentagonal rings, the same architectural principle as Fuller's geodesic domes. Pure carbon comes either as graphite or diamonds, so the discovery of a third, previously unknown form of carbon was remarkable. Colloquially known as Bucky-balls, fullerenes have such interesting chemical properties that the three discoverers were recognized with the 1996 Nobel Prize in Chemistry.

Fullerenes can be produced by lightning strikes, forest fires, and meteorite impact, but how long they persist in the environment or in rocks remains uncertain. Reported fullerenes in Cretaceous-Tertiary boundary clays have been linked to catastrophic wildfires triggered by the impact. The discovery of fullerenes in Japanese Permo-Triassic boundary sections was also linked to wildfires, with the fullerenes preserved in soot particles. "Trust but verify" is even more of a principle for geologists than for diplomats (or at least presidents) and scientists always want important results confirmed by independent laboratories. Unfortunately, confirming the geological reports of fullerenes has proven difficult. Many fullerene reports have not been replicated at other laboratories, and the reported fullerene abundances are often close to the limits of detection. One study found an increase in hydrocarbons, which may mimic fullerene, but no evidence for fullerenes themselves. Some chemists were hardly surprised by the failure to replicate early reports. Fullerenes break down relatively quickly when exposed to air. Preservation of fullerenes for hundreds of millions of years may be possible, but only under extraordinary conditions.

Fullerenes were reported from Permo-Triassic boundary sections at Meishan in China and in southwestern Japan,[12] evidently concentrated in the volcanic clay marking the extinction (see chapter 3). When Luann Becker and her colleagues analyzed the ratio of helium to argon within fullerenes, they found a surprise. Fullerenes were absent from rocks above and below the extinction point. Fullerenes can capture other elements within the bucky-ball structure; hence the great fondness chemists have for them. Those that arrive on meteorites have a very distinctive ratio of helium to argon, reflecting the primordial ratio in the planetary dust cloud during the formation of the solar system. This ratio is very different from the helium to argon ratio on earth. By comparing the ratio of helium to argon in fullerenes from Permo-Triassic boundary beds to fullerenes from the Murchison meteorite, Becker's group was able to determine whether the fullerenes were formed on Earth, perhaps in a wildfire, or came from space. The critical question was whether the fullerenes could have been produced during an impact rather than being remnants from the origin of the solar system. Fullerenes only incorporate gases like helium and argon into their structure when the surrounding pressure is very high, far beyond the pressures possible on Earth even during an impact, and must have formed in the planetary gas cloud and later have been delivered to Earth. They argue that this is convincing evidence for an impact and suggest an object about nine kilometers in diameter (although there seems little real evidence for the size of the object).

Such a novel approach to impact raises a host of questions: Are the assumptions about terrestrial versus extraterrestrial sources sound? Are fullerenes actually stable enough to survive an impact? Can fullerenes be carried along in the waters that flow through rocks, and even be preferentially enriched in certain zones? All of these questions had to be faced by the proponents of iridium in the early 1980s. It was some time before most geologists accepted an extraterrestrial origin for the high concentrations of iridium at the Cretaceous-Tertiary boundary. In the same way, although fullerenes have been studied in many different types of rock, it is probably still too early to declare fullerenes conclusive evidence of an impact at the Permo-Triassic boundary.

Just as I was finishing this book, new evidence of possible impact appeared, again from sites in Antarctica and China, in the form of microspherules composed of iron-nickel-silica and possible fragments of a meteorite. While the scent of an impact may be growing stronger, the stench of massive volcanic eruptions has become overwhelming, so we turn now to the second large class of proposed causes: volcanism.

· · ·

Mass extinctions and the Alvarez hypothesis have piqued the interest of many scientists far beyond paleontology and geology, and few months go by without my receiving an email bearing yet another theory for the end-Permian mass extinction. One of my favorite Grand Unified Permian Theories comes courtesy of several

Indian physicists. They linked the twin phases of the Permian mass extinctions at the end of the Middle Permian and the close of the Permian to a purported accumulation of weakly interacting massive particles (WIMPS) in the Earth's core. WIMPS are a form of hypothetical dark matter conjured up by some cosmologists to explain why galaxies have less observable mass than is required to explain their rate of rotation. If there is a great deal of hidden mass, everything will balance out and the cosmologists will be happy. (Personally, keeping cosmologists happy has never been high on my list of priorities, but I digress.) The Indian physicists proposed that an influx of high-energy WIMPS caused genetic damage and increased cancer rates in everything from plants to sponges to the therapsids of the Karoo. As the WIMPS accumulated in the core of the Earth they caused heating and eventually the eruption of a massive plume of very hot material from near the boundary between the core and mantle. Superplumes rise like a hot-air balloon through the mantle and eventually erupt at the surface where they may produce massive piles of volcanic rock. Thus cancer caused the first pulse of extinction and the eruption of the Siberian volcanism caused the second pulse (of which more, below). Changes in tectonics and anoxia in shallow marine waters completed the extinction process.[13]

Now I don't believe a word of this, but I do admire the sheer creativity: virtually all of the features of the two extinction pulses are incorporated, without worrying overly much about the details. Leaving aside the issue of why such hypothetical WIMPS should accumulate in the Earth's core, and how they would trigger a mantle plume, it would be easy to dismiss this scenario as a bit crazy. Much of this model is beyond the realm of science, but there are a few testable claims. Other elements are shared with other extinction scenarios and so do not provide any unique predictions. Physicists and geophysicists are better able to judge the plausibility of WIMPS heating the core and inducing plume formation, but the purported increased cancer rates during the Capitanian extinction are nicely undetectable in the fossil record. On biological grounds, mass extinction due to cancer is implausible and is unlikely to produce the different extinction patterns (why would some animals with a heavy shell, such as brachiopods, suffer considerable extinction, while molluscs with an equally heavy shell escape?). On average, one should expect higher cancer rates among the most exposed organisms. Thus terrestrial and shallow marine animals should have been the most affected. Species with large population sizes should be the most likely to survive. As we will see in chapters 5 and 6, different groups varied considerably in how much extinction they suffered, but the selectivity was not in the direction required by the WIMP hypothesis. Yet the WIMP hypothesis addresses two very real issues: was there a linkage between the two pulses of extinction in the latest Permian, and what triggered the massive outpourings of basalt in Siberia?

Until recently geologists were more comfortable with slow, inexorable processes rather than Las Vegas spectaculars, and the immense size of some flood basalts led geologists to expect that they formed over millions of years. Over about 1 million years an area about the size of the continental United States was inundated with at least 4 million cubic kilometers of material, covering the region to a depth of 100 to 6,000 meters or more. The environmental effects of so much volcanism, incessant clouds of dust, acidic vapors, and heat are difficult to contemplate today. We have no experience of eruptions on the scale of the Siberian flood basalts, the largest to erupt on any continent in the past 600 million years. The 1783 Laki eruption in Iceland generated 12 cubic kilometers of material in about eight months. But massive continental flood basalts seem to operate by different rules: a single flow from the Cretaceous-Tertiary Deccan flood basalt in India generated nearly 1,000 cubic kilometers in perhaps only a few weeks. Some volcanic ash and other debris are associated with the Siberian flood basalts so at least some of the eruptions must have been explosive, but most of the eruptions would have been much more like those of Kilauea in Hawaii than Mount Pinatubo in the Philippines, or Mt. St. Helens in Washington.

Basalt forms from a viscous magma that tends to flow during a flood basalt eruption, although some explosive eruptions have occurred. If basaltic eruptions are large enough, the magma will gradually fill in valleys and lowlands much as honey fills in the roughness of an English muffin. The flowing basalt obliterates the countryside, preserving it beneath a smooth, undulating surface, hence the name *flood basalts*. Flood basalts can cover a vast area with flows as much as a mile in thickness.

Today the main area of the Siberian flood basalt covers only about 675,000 square kilometers, but estimates of original extent continue to grow. Russian geologists have identified at least four distinct centers of volcanism. The Noril'sk region in northwest Siberia is one of the most easily accessible and thus well-studied zones. Here, volcanic material 3,700 meters thick includes eleven discrete eruptive sequences and forty-five separate flows. A single flow may be tens to hundreds of meters thick with layers of volcanic ash (tuff) and other volcanic debris several meters thick between the flows. One tuff layer 15 to 25 meters thick has been traced across 30,000 square kilometers. In another region, Maymecha-Kotuy, geologists have estimated the total thickness of volcanics at over 6,500 meters, or almost four miles. Erosion, burial, and the formation of the Ural Mountains have destroyed the western margin of the volcanics. The basalts also extend into central Kazakhstan. To the north remnants are found in the Taymir region, and kimberlite pipes to the east (the source of diamonds) represent an early, very explosive phase of the eruptions.[14] The total region of volcanic material covers much of Siberia from the Urals east to Lake Baikal and south into Kazakhstan (figure 2.6), an area of about 7 million square kilometers (2.7 million square miles)—almost equal to that of the continental United States.

Volcanic eruptions incorporate radioactive elements that decay over time, as discussed in more detail in chapter 4. This decay produces a sort of radiometric clock that allows the date of the eruption to be calculated. Recent dating of the Siberian flood basalts has produced ages ranging between 252.2 and 251.1 million years ago, indicating the flood basalt erupted in 1 million years or less. These dates are essentially identical with other, high-quality radiometric studies of Permo-Triassic volcanic ash beds in south China, suggesting the events were contemporaneous.

For the moment, let us assume that the dates for the flood basalts and the mass extinction overlap, occurring during the same mil-

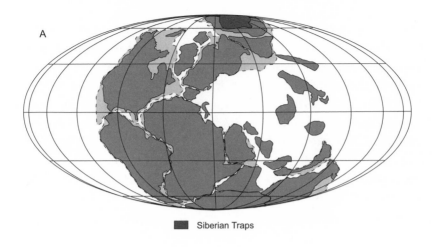

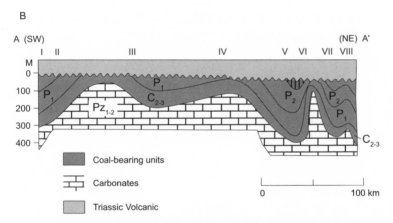

Figure 2.6 The Siberian flood basalt. A: reconstruction of Pangaea at the Permo-Triassic boundary showing the position and extent of the Siberian volcanism. B: a generalized cross-section across Siberia showing the coal-bearing units beneath the volcanics. A is based on Scotese and Langton (1995); Ziegler and Charles (1990); and Ziegler et al. (1998); B is based on Czemanske et al. (1998), reproduced with permission from *International Geological Review* 40(2): 95–115, © V. H. Winston and Son.

lion-year period. Coincidence? Perhaps, but a pretty unusual one. Coincidences plague any historical analysis and the correlation of two events in time does not prove a causal connection between them. The human genome was deciphered the same year George Bush the younger was declared president, but it is hard to see any connection between the two events. What we really want to know is how the flood basalt eruption could lead to the extinction of so many species, a subject to which we will return in chapter 8.

Ash beds near the Permo-Triassic boundary in south China reflect a different style of volcanism. My colleague Professor Yin
Hongfu of the China University of Geosciences at Wuhan has long
led one of the major Chinese groups studying the extinction.
There are many volcanic ash beds in the Chinese sections spanning the Permo-Triassic boundary; a particularly nice ash bed coincides with the mass extinction. Unlike the flowing flood basalts,
volcanoes along plate margins where oceanic crust is being subducted into the earth behave very differently. The subducted slab
melts and the large amount of silica produces a thick, gummy
magma that often erupts explosively. Subduction-derived basalts
also have distinctive chemical signatures. From this we know that
the south China ash beds were not related to the Siberian flood
basalts, but to volcanoes along the southern margin of China.
Even the small amount of ash and gas produced from the eruption
of Mt. Pinatubo showed that stratospheric winds distribute ash
globally. Some of the latest Permian volcanic eruptions in south
China were far larger than Mt. Pinatubo, or any in recorded
human history. This led Professor Yin to propose that the eruptions blotted out the Sun, cut off photosynthesis, and triggered the
mass extinction. But there are many massive volcanic eruptions
documented in the rocks of south China, and only one corresponds with any loss of species.

· · ·

Continental drift and its impact on the earth's biota through climatic change form the third class of extinction hypotheses. Despite having spent my teenage summers in the incredible rocks
of the southwestern United States, I entered college woefully ignorant of geology. This deficit was quickly mended through a
freshman seminar on plate tectonics: the horizontal movement
of plates that comprise the upper 10–40 km of the earth, their
creation along midocean spreading centers and their destruction
as they plunge into the earth and gradually melt. That first course
also provided my introduction to the end-Permian mass extinction, when I discovered the work of Jim Valentine relating plate
tectonics and changing positions of the continents to changes in
biological diversity.[15] Ecologists had long understood that one

factor controlling the total number of plants and animals on the globe was the number of biologically independent regions, or biotic provinces. All other things being equal (not that they ever are), the more biotic provinces, the greater total global biodiversity. Jim's insight was to realize that the dance of the continents should play a profound role in modulating biodiversity over millions and tens of millions of years by changing the number of provinces and climatic patterns.

Imagine that three continents collide, each containing separate species of rabbits. Before the collision each rabbit species has a continent to itself, but after the collision the rabbits must compete with each other. One expectation from ecology is that only the best-adapted rabbits will survive on the new, larger continent. The other two rabbit species will either become extinct or find a different way of making a living. If the continents are large enough, there may be many species of rabbits in different habitats, but the same logic applies. This is, incidentally, the reason that conservationists are so concerned about the spread of non-native species around the world: such biotic homogenization reduces biotic provinciality just as surely, and far more rapidly, than continental drift. Conversely, the breakup and dispersion of continents should increase biological diversity as the species on each continent follow independent evolutionary trajectories.

Continents may drift into different climatic regions but they can also modify climate in less obvious ways, for example by changing oceanic currents. As the Isthmus of Panama formed a few million years ago it closed off a westward equatorial current, turning the Caribbean into a large gulf. The currents that once flowed through the Isthmus were forced to turn north, considerably strengthening the Gulf Current. Great Britain and Europe have greatly benefited as this northward direction of warm water into the Atlantic has made them far more habitable than would otherwise be the case. Jim Valentine reasonably concluded that such continental movements must be important controls on long-term biodiversity patterns.

When Jim was considering these issues during the early 1970s, reconstructions of past continental positions were fairly primitive.

The revolution in plate tectonics was less than ten years old and the supercontinent of Pangaea was believed to have formed near the end of the Permian (as shown in figure 2.6). Pangaea was an agglomeration of almost all the continents, formed by a collision between a southern mass of continents (Antarctica, Africa, South America, India, Australia, and bits of southeast Asia) known to geologists as Gondwana, and a northern mass (North America, Europe and Siberia), or Laurasia. The formation of Pangaea suggested to Jim and his colleague Eldredge Moores that a reduction in biotic provinciality caused the mass extinction. In essence, as Pangaea grew the homogenization of previously distinctive plants and animals drastically reduced biodiversity.

The formation of such a large landmass would influence global climate, which might also have contributed to the extinction. Siberia and Chicago have more severe climates than New York or London: the winters are colder and more miserable, and the summers warmer and more humid. This greater seasonality in continental interiors is a simple consequence of the ameliorating effects of the oceans. More energy is required to change the temperature of a given volume of water than the same volume of air. Thus oceans moderate the climatic fluctuations of coastal areas relative to the interior. Large lakes, incidentally, can have a similar effect (so Chicago would be even more miserable in February without Lake Michigan). Central Asia has very pronounced fluctuations in climate today, with cold winters and very hot summers, but these variations are mild compared to estimates for the center of Pangaea. Since we know from fossils that plants and animals did survive across Pangaea, large lakes may have existed in central Pangaea to modify the climate.

The increasing seasonality caused by the formation of Pangaea should also affect animals along the coastline. By impoverishing the ecosystems and changing current and climate patterns, Valentine and Moores suggested the supply of nutrients might have become more uneven, possibly leading to the disappearance of groups with very specialized nutrient requirements. Generalized species capable of feasting on whatever resources were available would increase. By this model, the formation of Pangaea should have caused a loss of biotic provinces, an increase in climatic variability, and an increase in the variability of nutrients, all contributing to the mass extinction.

. . .

One of the most vexing features of the Permo-Triassic boundary is how hard it is to find. Permian and Triassic marine rocks are not uncommon in many parts of the world, yet few rocks were deposited during the transition between the two. This paucity of marine rocks was evident in the late 1960s and led Norman Newell of the American Museum of Natural History, one of the great paleontologists of the twentieth century and a founder of the more biological approach to paleontology, to invoke a global drop in sea level as the cause of the extinction, the fourth group of extinction scenarios. Why a drop in sea level should cause a mass extinction at first seemed obscure; Newell's paper and earlier work had suggested that regression could cause extinction by crowding species into less space. In 1967 Robert MacArthur and E. O. Wilson penned a short book that has become one of the classics of ecology: *A Theory of Island Biogeography.*[16] The isolation of islands makes them ideal systems for studying ecology, including the processes that control species diversity. MacArthur and Wilson argued that the number of species on an island reflected a balance between the immigration of species from the mainland, and the loss of species on the island due to competition with other species. From this it followed that small islands and those far from the mainland should have fewer species. Converting the model to evolutionary time replaces immigration with speciation and suggests that the species diversity on islands (and by extension other areas) should reflect a balance between speciation and extinction. The balance between these processes will be determined by the size of the island.

Extending this theory into the fossil record implies that since fluctuations in sea level alter the area of the shallow marine continental shelves occupied by most marine plants and animals, a drop in sea level should reduce shelf area, increase competition, and cause species extinction. During the 1970s many biologists believed that the relationship was nearly linear, so that the change in species diversity would correspond to changes in area. If, as many geologists argued, the end-Permian drop in sea level was so great that it virtually completely eliminated the deposition of marine rocks anywhere in the world, then a significant mass extinction was almost inevitable,[17] although predicting the size of the expected extinction turns out to be rather more difficult.

In principle, it should be straightforward to take the change in the area of shallow seas caused by a drop in sea level and determine how many species should become extinct. But while geologists could gauge the vertical extent of the change in sea level (with estimates varying between 200 and 280 meters), translating this into area is far more difficult. The change in area depends on how much of the continents are flooded and the slope of the continental shelves. The ice ages of the past few million years generated sea level changes of hundreds of meters but relatively little extinction because the oceans covered so little of the continental shelves: extensive changes in sea level were not translated into much actual change in area. Moving to the Late Permian, the efficacy of this model depends on the slope of the continents during the Late Permian, something we cannot readily determine.

If we assume for the moment that the estimated drop in sea level is correct, another approach to understanding the extinction is to ask what could trigger such a significant drop in sea level. Although sea level rises and falls for a variety of reasons, and on a variety of time scales, the change at the end of the Permian appeared to be one of the largest in the past 600 million years. Moreover, this drop is followed by a very fast rise in sea level, one of the most rapid known. This pattern suggested to Steve Stanley of Johns Hopkins that the extinction and the drop in sea level could be explained by a widespread continental glaciation.[18] Glaciation causes extinction as water is removed from the oceans to form ice, reducing the amount of shallow continental shelf, and as plants and animals have to migrate toward the equator to escape glacial conditions. Species unable to migrate may perish. Stanley cited other geologic factors to support this idea, including Late Permian glacial debris found in Siberia and in eastern Australia, the extinction of tropical reefs, which commonly suffer during glaciations, and the paucity of limestone deposition. Glaciation appears to have been the primary cause of the second largest mass extinction at the close of the Ordovician, lending credibility to the suggestion. There is little evidence for Late Permian glaciation sufficient to cause the estimated drop in sea level, raising doubts about the role of glaciation in the extinction.

· · ·

By the early 1990s several geologists began to question the accepted wisdom that sea level had dropped at the end-Permian extinction. Tony Hallam and Paul Wignall have most fully developed this view based on Hallam's earlier work on the end-Triassic mass extinction.[19] They trekked through China, Pakistan, the Alps, and elsewhere, carefully analyzing the changes in fossils and the types of rock formed during the extinction. These studies considered relatively few species and did not analyze the data statistically, but their data were convincing. The pattern of rock deposition revealed that sea level had been rising—challenging the views of pretty much every other geologist working on this issue.

But how does rising sea level kill things? This reversal in perspective leads to the fifth group of extinction scenarios, a cluster of three hypotheses all invoking a drop in levels of oxygen in the ocean, or anoxia. Hallam and Wignall focused on the rocks deposited just above the Permo-Triassic boundary. Earliest Triassic communities have few species, although the numbers of individuals in any one species may be enormous, running to millions or tens of millions of individuals. Such a skewed distribution is typical of highly stressed environments or postdisturbance settings when opportunistic, weedy species dominate. Indeed some of the fossils represent just such opportunistic groups. *Claraia*, first introduced in chapter 1, is a very thin scallop commonly found in low-oxygen waters, as is the brachiopod *Lingula*, another common species in earliest Triassic rocks. The very thin, rhythmic beds of Early Triassic limestones often are produced in environments where no worms and other burrowers were present to burrow through the mud, mixing the sediment and destroying fine laminae. The rocks also have peculiar chemical patterns that convinced Hallam and

Wignall that rising sea levels brought very low oxygen, or dysaerobic, waters into shallow water. Most marine animals are adapted
to normal oxygen levels, and if the amount of oxygen in the water
drops too low, they essentially suffocate. Since the low oxygen waters will reach localities far out on the continental shelf before
the waters reach farther inland, we should be able to trace the
migration of the extinction with the rising, anoxic waters.

Anoxia figures in a different extinction model proposed by my
colleague at the University of Tokyo, Yukio Isozaki, based on
fieldwork in Japan and British Columbia. Sections across the
Permo-Triassic boundary in Japan represent scarce bits of deep
ocean sediments, and Yukio has focused his research on events in
the deep sea. The recycling of plate tectonics destroys most oceanic crust, but as an oceanic plate is driven beneath a continent,
slivers of the oceanic plate are often smeared onto the side of a
continent. (The technical term for this is *obducted*, to contrast with
subducted. Wonderful word, *obducted*.) Particularly common in
these slivers are deep-sea cherts produced from the constant rain
of siliceous microfossils. (Quartz is a highly crystalline form of silica, while chert and agate are microcrystalline silica, while glass
and obsidian are amorphous or noncrystalline silica.) When deep
ocean waters and sediments contain some oxygen, the resulting
chert is a dark brick red as iron reacts with the oxygen. Above and
below the Permo-Triassic boundary the cherts are dark red, but
the boundary itself lies in dark gray to black cherts and carbon-
rich claystones, signifying a lack of oxygen. The Pacific was actually
much wider in the Permian than today (since the Atlantic didn't
exist), and the discovery of similar rocks in Japan and British Columbia suggests that the entire deep ocean may have been anoxic
for perhaps 20 million years.[20]

An anoxic deep ocean is a very odd thing to contemplate. Today
upwelling currents bring nutrients from the deep sea into shallow
waters, producing some of the richest ecosystems in the oceans.
Other currents form as cold water sinks near the poles and drives
toward the equator. Together these processes mix the ocean and
ensure some oxygen even in the deepest levels. If Yukio is correct,
no mixing occurred in deep oceans across the Permo-Triassic tran-

sition. Instead, the oceans must have been stratified, with an oxygen-rich surface ocean isolated from an oxygen-poor deep ocean. In Isozaki's extinction model, the stratified ocean develops near the end of the Middle Permian, triggering the first pulse of extinction, and then reaches a peak at the Permo-Triassic mass extinction, before gradually dying away in the Early Triassic. Thus this single hypothesis links two pulses of extinction and the long delay in the recovery of biodiversity after the mass extinction to a common cause. Critically, the deep-water persisted for perhaps as long as 10 million years.

Just as medicine has developed CAT scans, cardiac stress tests, and all manner of noninvasive imagery to measure the relative health of our bodies, geologists have developed a range of tools to chronicle the health of the earth through time. Identifying changes in the flow of carbon through the earth, oceans, and atmosphere is a powerful tool for tracking changes in the health of the earth. The end-Permian mass extinction coincides with an abrupt shift in carbon isotopes, evidence of a massive change in the carbon cycle. How we study these changes will be discussed in more detail in chapter 7, but a brief disquisition may help to explain the next two models.

Carbon comes in several different isotopes. Each isotope has the same number of protons in the nucleus but differs in the number of neutrons. Some isotopes, including carbon-14, are unstable and spontaneously decay to more stable isotopes (in this case nitrogen-14). During photosynthesis, the lighter isotope of carbon, carbon-12, is taken up in preference to carbon-13, so plants and the animals that feed off them have more carbon-12 than the average ratio in the atmosphere. This enrichment of organic material in carbon-12 continues into any organic material resulting from the decay of plants and animals, including humus and peat in soils, oil and organic material in rocks. This produces two large reservoirs of carbon differing in their ratio of carbon-12 to carbon-13: an inorganic reservoir and an organic reservoir enriched in carbon-12. Photosynthetic activity in shallow marine waters preferentially removes carbon-12 leaving the surrounding waters with a greater amount of carbon-13. Shells and skeletons record the ratio of the waters when they are formed, so by analyzing the ratio of carbon-12 to carbon-13 in a series of fossils, geologists can track changes in this ratio.

If that was all there was to the carbon cycle, the National Science Foundation would not have spent tens of millions of dollars over the past few decades equipping geochemists with the latest in spiffy new mass spectrometers to measure these isotopic changes. In fact, the changing ratio of carbon-12 to carbon-13 is a very sensitive indicator of changes in the carbon cycle. Comparing the carbon ratio between shallow waters and deep waters across the Cretaceous/ Tertiary boundary shows, for example, that photosynthesis virtually disappeared during the mass extinction. The changing ratios may also reveal major changes in the amount of carbon added or removed from the organic carbon reservoir. As discussed in chapter 7, with a few assumptions geochemists can determine the amount of carbon that has shifted between one reservoir and another and from this the most likely sources of carbon.

The abrupt shift in the ratio of carbon-12 to carbon-13 at the Permo-Triassic boundary suggests that a large volume of organic carbon was added to the oceans and atmosphere (as I will discuss later, the rate of burial of organic carbon might also have dropped). But the curse of carbon isotopes is that many different processes can produce the same pattern. Give a bunch of PhDs data and few limits on their creativity, and something akin to chaos is the unremarkable result. Consequently, many of the disagreements over the cause of the mass extinction are really a debate over the cause of this shift in carbon isotopes. The following two ideas are based almost exclusively on differing views of the carbon isotopic evidence.

Some of my most significant contributions to science reflect my own stupidity. In 1994 I was at a meeting with Andy Knoll of Harvard and John Grotzinger of MIT. I showed a slide of a stromatolite (a rock with many very thin layers produced by microbial activity). John nudged Andy and said, "That's not a stromatolite, it's an inorganic precipitate." I would choose the wrong slide to show before the only two people who could tell them apart. Following my talk they descended on me like a pair of hungry wolves (well, to the extent that a couple of distinguished Harvard and MIT professors can act like hungry wolves). John and Andy wanted to know where I had taken the photo and got very interested when I told them that it was from near the top of Guadalupe Peak, the highest point on the Permian Reef in West Texas. These sediments may have been laid down during the earlier, Guadalupian mass extinction.

In his work on 2-billion-year-old rocks from Canada, John demonstrated that stromatolites could form directly through precipitation of carbonate from the ocean, not just by microbial activity, as most paleontologists believed. The unusual chemical conditions required for this were common a billion or more years ago, but are much less common in younger rocks. The precipitates, the carbon isotopic shift, and a new perspective on the patterns of extinction and survival among marine species led Andy, John, and their colleagues to a very novel stance on the extinction.[21] They made two points: First, that a detailed analysis of the patterns of extinction and survival suggest that more metabolically active species were more likely to survive than those groups that lacked such active metabolism. What does metabolically active mean? Humans have a higher rate of energy use than sloths; thus they are more metabolically active. In the oceans, clams, snails, and crabs, for example, are more metabolically active than corals or brachiopods. Second, they proposed that carbon dioxide poisoning was the trigger for extinction. In other words, a vast amount of carbon dioxide was released into the oceans and atmosphere and essentially poisoned the animals. The source of carbon dioxide is difficult to explain, but the deep-water anoxia described by Yuchio Isozaki provided a clue. They followed Isozaki's claim that the deep oceans had been stagnant during the Late Permian, and proposed the buildup of massive volumes of carbon in these stagnant, deep waters. Such a stagnant ocean is inherently unstable, and eventually something will cause it to overturn, releasing the carbon as carbon dioxide. In their model the onset of glaciation shifted current patterns in the oceans, triggering the release. Following their 1996 report, the *New York Times* drew a bad analogy between this model and the great burp of carbon dioxide from Lake Nyos in Cameroon on 26 August 1986 that killed about 1,700 people. In fact, Andy and colleagues did not argue for such a convulsive burst, but more for a long slow fizz, with the mode of extinction carbon-dioxide poisoning rather than asphyxiation, as in Lake Nyos.

. . .

It took me quite a while to understand the data on carbon isotopes from the Permo-Triassic boundary, and what little understanding I do have is thanks to the late Bill Holser, a geochemist at the University of Oregon who patiently tried to explain them to me during the early 1990s. As Bill's teaching began to sink in, I realized that none of the then current extinction models adequately explained the shift in carbon. This leads to the sixth extinction model. Picking up on suggestions by Euan Nisbet of the University of Edinburgh and others, in 1993 I proposed that the drop in sea level could have triggered the release of a large amount of methane from sediments on the outer continental shelf. Some bacteria produce methane rather than oxygen as a by-product of feeding off organic material, and in the ocean below about 300–500 meters methane can build up in the sediment as a sort of slush. The high pressure causes water to freeze at higher temperatures, but this produces cavities in the ice. These cavities fill with methane, producing methane hydrates. Oceanographers have become quite fond of them because the ices burn when touched by a flame, producing no end of amusement. The methane is stable in deep sediments because the pressure of the overlying water stabilizes them, but if the pressure drops, either because sea level drops or because some one punches a drill core into the ocean bottom, methane can be released to the water and eventually the atmosphere. Methane is a much more powerful greenhouse gas than carbon dioxide but in the atmosphere it normally dissociates to carbon dioxide in ten to twenty years. Nisbet suggested that the release of methane hydrates as sea level dropped during the recent glaciations reversed the climatic cooling, warming the globe and ending the glacial phase.

It took no particular insight on my part to realize that if major drops in sea level during the ice ages had released methane, the same thing was likely to have happened during the end-Permian sea level regression. The drop in sea level leads to release of gas hydrates, triggering the shift in carbon isotopes and producing global warming and extinction, but I was never entirely convinced that methane release was sufficient to explain the entire extinction.

To summarize the various hypotheses, we first have the possibility of the impact of an extraterrestrial object with the earth, as in the end-Cretaceous mass extinction. The second hypothesis involves the climatic consequences of the massive flood basalt volcanism in Siberia, and suggestions that the volcanism was somehow caused by an impact. Third, the formation of the supercontinent of Pangaea during the Permian may have caused biotic homogenization, just as is happening today, and gradually reduced biodiversity. The formation of Pangaea is one of the possible explanations for a drop in sea level at the Permian-Triassic boundary, the focus of the fourth group of extinction hypotheses. Here, a reduction in the area of shallow seas may have caused extinction, and Stanley suggested this could have been caused by a glaciation. Fifth is a set of three independent hypotheses invoking oceanic anoxia. Wignall and Hallam argue from detailed field studies that any drop in sea level was over by the extinction, and rising seas brought anoxic waters into shallow seas. Isozaki is more concerned with the growth of anoxic waters in the deep ocean, which Grotzinger, Knoll, and Bambach employed in their model of carbon dioxide poisoning. Finally, the sixth model was my suggestion that the release of vast amounts of methane from the continental shelves may have caused global warming and other climatic effects, leading to the extinction.

Each of the theories outlined above proposes a single primary cause for the mass extinction. Subsidiary causes trigger the extinctions, but each flows from the primary trigger. Life would be much easier if complex events had single causes, but the lessons of history are otherwise. The causes of most complex historical events are notoriously difficult to pin down, and the same is likely to be true of events in the history of life. In 1993 I suggested that methane release combined with several independent factors to trigger the Permo-Triassic extinction. I christened this the *Murder on the Orient Express* hypothesis, after the Agatha Christie murder mystery where all the suspects perpetrated the crime (although in Dame

Agatha's novel the victim deserved his fate, while the brachiopods most assuredly did not). The chief drawback of this proposal is that it is very difficult to test, leading a colleague of mine to call it "Erwin's kitchen sink hypothesis" (he was not being complimentary). The difficulties in testing such ideas are not a priori a reason for dismissing them.

. . .

Scientists have done a great marketing job convincing the public that we are objective seekers of truth, dispassionately weighing the evidence in our undying and noble quest to understand how the world works. Mr. Spock from *Star Trek*, but with a beard. The reality is that scientists are often opinionated, egotistical mavericks trying to determine how the world works while generally having a bit of fun along the way. Like many other people some scientists are also sure that we are right and anyone who disagrees with us is wrong. This is not to say that scientists do not change their minds. We do, if only under duress. Some of the best scientists are most willing to change their minds. But research is often a battlefield between contending egos, uncomfortably constrained by reality. Fortunately for the enterprise, scientists as a community have generally agreed on a series of methods to explore that reality, and ultimately it is this evidence that allows us to test differing views about how the world works. In the remainder of this chapter I summarize the explicit tests that allow us to evaluate these hypotheses.

Testing the impact proposal for the end-Permian mass extinction is easy, at least in principle. In fact, dozens of such tests have been conducted in the past two decades, and most have failed, with the possible exception of the recent discovery of fullerenes, meteorite fragments, and microspherules. Research on the Cretaceous-Tertiary boundary suggested we should search for such signs of impact as increases in the abundance of specific elements, particularly iridium, shocked quartz, diagnostic sedimentary deposits formed by the material ejected from the crater, and possibly fullerenes and other impact debris. The search for all of these features will continue, with increased attention placed on testing the reliability of the fullerenes. The work of Luann Becker and colleagues on fullerenes will have to be replicated by other labs, and similar evidence sought at other boundary localities. Geologists will also want to learn more about the origin of the fullerenes, how they may move in sediment, and if the helium and argon evidently encased in the fullerenes is truly an extraterrestrial signal. New investigations of the Bedout structure off Australia will be needed to test suggestions that it is an impact structure, and provide reliable estimates of its age; early reports are not encouraging. Furthermore, if the extinction was due to an impact, the event must have been virtually instantaneous and global, involving land and sea.

Three primary questions face proponents of a connection between the Siberian flood basalt and the mass extinction: First, do the radiometric ages for the PT boundary and the flood basalt really overlap? As these ages are refined, the correlation may fade and with it the link to the mass extinction. Second, what caused the eruption of the flood basalt? Greater understanding of how flood basalts form is essential to testing the suggestion that a massive extraterrestrial object triggered the eruptions. Understanding their source is not critical to establishing a link to the mass extinction but may reveal whether an underlying cause was present. Finally, there remains the issue of the linkage between the eruption and the extinction. This is the most difficult question, and parts of it may be impossible to resolve, but if the correlation between the flood basalt and the mass extinction withstands further scrutiny, the causal connection between them demands attention. Correlation is not causality.

Perhaps one way to examine the connection between massive volcanic eruptions and the extinction is by comparing the effects of volcanic eruptions of a similar size to the eruptions at the Permo-Triassic boundary. The South China volcanism is of similar magnitude to several well-studied volcanic eruptions in the past 100 million years. In addition, ash beds from many other large eruptions are preserved in Late Permian and Early Triassic rocks in south China, providing an opportunity to examine the fossil record above and below them. If volcanic eruptions were a major factor in the mass extinction, we would expect to see a stepwise extinction, with each step at an ash bed.

The Valentine-Moore hypothesis requires that the extinction occurred over millions of years as Pangaea formed. The pace of the extinction must match the slow drift of the continents, and the pattern of extinction must correspond to the onset of collision between continental regions. With an adequate fossil record, paleontologists should be able to track the intermingling of faunas and resulting extinctions. One test of this model is to determine whether the mass extinction and the formation of Pangaea happened at the same time, or whether the mass extinction occurred more rapidly than the slow movement of the continents.

A number of tests for the species-area effect are available. This model requires that the drop in sea level was contemporaneous with the extinction. If they are out of phase, particularly if sea level is rising rather than falling, the hypothesis fails. Other information is available to test the model by looking at the effects on biodiversity of other changes in sea level and through recent studies of the species-area effect.

As continental glaciers grow water is removed from the oceans, reducing sea level. If glaciation were responsible for the mass extinction the drop in sea level should correlate with the peak of extinction, and with geological evidence for glaciation near the poles. We should also expect that the rate of the regression and subsequent rise in sea level would be similar to other glacial sea-level changes. If changes in sea level do not correlate with the extinction, or if the change in sea level is significantly slower or faster than is typical of glacially induced changes, we could reject this model.

The various anoxia models make very specific predictions about the nature of the extinction. Both the Hallam-Wignall transgression model and Isozaki's model focus on marine extinction. The Hallam and Wignall model requires a tight connection between the appearance of geological indicators of low-oxygen waters and the extinction, with the extinction occurring progressively as sea level rises. In Isozaki's deep ocean anoxia model, the actual mode of extinction is poorly articulated, but is evidently associated with the spread of the stagnant oceans. This model would be falsified if the extinction occurred on a different time scale than the anoxia, or if diverse faunas occurred in shallow waters at the same time as the super-anoxia was present in the deep oceans.

The oceanic overturn model of Andy Knoll and colleagues involves two partially independent hypotheses; each must be tested separately. First, the differences in extinction and survival should reflect species differences in ability to withstand high levels of carbon dioxide. Second, in their model carbon dioxide was produced from a reservoir in the deep sea, where it had built up over millions of years. Glaciation induced more vigorous circulation (the opposite of the claims of Isozaki's model) and initiated the release of the carbon dioxide. But the increased levels of carbon dioxide could be from a different source, and thus falsifying the mechanism does not necessarily invalidate the model.

The *Murder on the Orient Express* hypothesis is the most difficult to test. We may discover, however, that we can eliminate a number of hypotheses, but are left with several that are equally consistent with the available data. If so, two questions arise: Is there data that geologists could collect that would allow us, in principle, to discriminate between these hypotheses? Is there some underlying single cause that could unite all this seemingly distinct data? Particularly if the answer to the second question is no, we would have to seriously entertain the idea of multiple causes, perhaps interacting in a complex fashion.

Discriminating between so many possible causes is far simpler than it appears. As we move through subsequent chapters, keep the following issues in mind: Was the extinction rapid, even catastrophic, or a slower, more drawn-out event? This will clearly distinguish between several proposals. How well do environmental and climatic events correlate with the episodes of extinction? This can help separate cause and effect. Do the patterns of extinction match those expected from various extinction scenarios, in particular in the anticipated involvement of marine and terrestrial realms?

The distribution of geological resources across the globe is highly uneven: oil in the Mideast, water in some areas and deserts in others, and Permo-Triassic boundary rocks in China. Rocks spanning the boundary are found in other places: among them Greenland, Pakistan, Iran, and northern Italy. But these sequences are so plentiful in south China that any search for the causes of the extinction must begin there.

Meishan, Zhejiang Province, China
October 2003

I am teetering precariously on a nubbin of rock halfway up the cliff of an abandoned phosphate mine. Out in the valley the red tile roofs of Meishan provide a vaguely Mediterranean air to the subtropical setting of south China. The morning rain and fog have lifted, and through the haze, out beyond the village and a small river, I can see piles of coal and the mine adits that deliver miners into the warren of tunnels deep below me. *Meishan* means coal mountain in Chinese, and thick deposits of coal lie several hundred meters under the surface, deposited by the Middle Permian peat swamps that once covered this area. The seas gradually rose after the peat was laid down, covering the deposits with lime muds of the later Permian.

Off to my right I can hear the chatter of the rest of our group, and the sounds of a hammer as someone collects samples. I turn back to the cliff, continuing in my inept imitation of a mountain goat. My feet are planted on the very end of the Permian, a foot-thick layer of limestone known to boundary aficionados as bed 24. The dark gray limestone is stained a rusty red at the top. A close look at the rock reveals fragments of brachiopod shells and other

Figure 3.1 The Permo-Triassic boundary at Meishan, south China. The reddish stain tops bed 24, with the last diverse Permian fossils. The white clay is bed 25, and is topped by the black material of bed 26. Both beds 25 and 26 are volcanic clays, although possibly with additional components.

fossil debris. We cleared off this shelf to collect samples of beds 25 and 26, inch-thick layers of volcanic clay that mark the extinction itself. The clays are soft and pliable despite their great age, and after removing the overlying rock, we can carve them out with a putty knife (figure 3.1). Above bed 26 is another couple of feet of thick limestone, which gives way to a much thinner bedded unit devoid of fossils except for clams and occasional ammonoids.

I am back at Meishan with my close friends and colleagues Sam Bowring of the Massachusetts Institute of Technology, one of the premier age-daters or geochronologists in the world, and Jin Yugan of the Nanjing Institute of Geology and Palaeontology, an authority on brachiopods and the Permo-Triassic boundary (figure 3.2). I first visited Meishan with Lao Jin in 1992 and have visited many times since.

Beyond wandering up the Hong Sui River in Guangxi, and climbing through bean fields outside of Chongxing, over the past decade I have scoured gullies in a remote corner of northern Sichuan, and carried backpacks of ash samples through an active quarry near the Yangtze River as dynamite exploded around me. Many of the fundamental advances in understanding the events of the Permo-Triassic transition stem from research in south China. Early work on the Permo-Triassic mass extinction during the 1960s and 1970s by Western geologists was conducted in ignorance of Chinese geology. Although Western geologists began visiting China soon after President Nixon's opening to China in the early 1970s, it was not until a decade later that the wealth of Chinese data on the Permo-Triassic boundary became apparent. Chinese geologists have identified at least fifty different marine Permo-Triassic boundary sections (primarily by Jin's group at Nanjing and Professor Yin Hongfu's group at Wuhan), and many have been studied in detail. As we will see, from China we have received a means to correlate rocks around much of the world and this allows us to place events in the correct sequence. In China we were also able to establish that the extinction was far more rapid than anyone had imagined, and that the change in the carbon cycle was intimately tied to the extinction. The next chapter continues the story, describing how the volcanic ash beds in China, when analyzed in Sam Bowring's lab at MIT, revealed the pace of the extinction.

Through all of these trips I have returned to the Meishan quarry in Zhejiang Province because it holds so many secrets to the mass extinction (figure 3.3). The rocks at Meishan are one of the best-studied Permo-Triassic boundary sequences in the world and so have played a pivotal role in understanding the extinction. Indeed, this sequence of rocks is so well analyzed that it has been ratified as the global reference section for the Permo-Triassic boundary, what geologists call a Global Stratotype Section and Point (GSSPs, in the argot). Such reference sections provide a specific locality and a point in rock to define the boundary between two stratigraphic intervals. Rocks spanning the boundary elsewhere in the world are referred to this section through a variety of correlation tools, from the use of fossils in biostratigraphy, familiar from Murchison's work mentioned in the preceding chapter, to more recent means. The development of such boundary sections is only a few decades old. Boundaries were previously of little concern to geologists, and were generally defined by the

Figure 3.2 Professor Jin Yugan at a locality in south China, photographing the Late Permian ammonoid shown in figure 3.7. The boy and his father looking on are local fishermen who Jin and I hired to take us up the Hongshui River.

section where they were first recognized and described. A growing interest in the biologic and geologic crises that often accompany the major boundaries has spurred a more rigorous effort to define the boundaries. Geopoliticians have also appeared, wandering the world in learned discussion (and occasional nasty disputes driven as much by regional pride as science) over just where to place the various GSSPs.

Testing ideas about the cause of the extinction requires having an actual record of the events associated with the extinction. Meishan and other sections of rock spanning the Permo-Triassic boundary in south China have taken on such significance because they preserve just such a record. But in much of the world this is impossible to find. The rock just is not there. Recall the gap I described in chapter 1 between the late Middle Permian rocks in Texas and the Early Triassic rocks of Utah. Permian rocks provide the limestones capping the rim of the Grand Canyon in Arizona, make up the Guadalupe Mountains of the Texas–New Mexico border, and surround the Sange de Cristo Mountains near Santa Fe. Thousands of square miles of Early Triassic–age

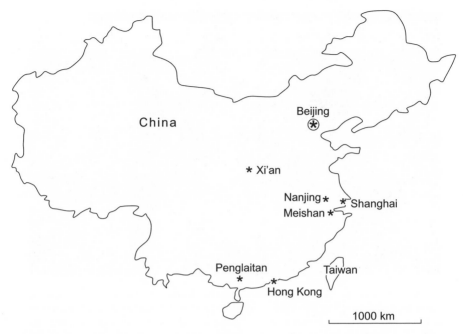

Figure 3.3 Location of the critical Meishan Permo-Triassic boundary section in south China, south of the Yangtze River, between Shanghai and Nanjing. The Penglaitan boundary section discussed in chapter 4 is shown near Hong Kong.

rock are piled right on top, extending through Utah into Montana and Idaho. How hard could it be to find rocks recording the transition from the Permian to the Triassic? But the boundary is missing, swallowed by a gap millions of years long. So either rocks spanning the Permo-Triassic boundary were never deposited in this area, or they were deposited and eroded before the Triassic rocks were formed.

The same problem occurs in many other parts of the world. Nicely preserved Permian rocks lie just beneath fine Early Triassic deposits in Greece, Sicily, and many other places, but just as in Arizona, there are millions of years of time missing at the contact between the Permian and Triassic rocks. Lest the conspiracy theorists get too excited however (I can see the headlines: "Extinction Record Stolen by Space Invaders!"), such unconformities are common throughout the rock record. These gaps at the Permo-Triassic boundary led geologists to the logical conclusion that little ma-

Figure 3.4 A larger view of the Meishan Permo-Triassic boundary sequence, showing beds 24–27 at Meishan. The last diverse Permian fossils are found in bed 24. Beds 25 and 26 comprise the volcanic ash beds that coincide with the mass extinction. The formally defined Permo-Triassic boundary actually occurs near the middle of bed 27c.

rine rock was deposited, because sea level dropped dramatically. Fortunately the rocks were not missing in China.

• • •

Meishan is a half-hour's drive outside of Changhsing City in Zhejiang Province and just midway between Shanghai and Nanjing. The latest Permian rocks at Meishan belong to the Changhsingian Formation, a series of limestones and cherts that were deposited in fairly deep water. The limestones are dark gray to black, reflecting the relatively high amount of organic material in them. The last bed with abundant Permian fossils is bed 24, a fairly thick unit of lime mud now turned to rock (figure 3.4). Beds 25 and 26 are volcanic ashes, now turned to clay. Bed 25 is light gray clay while bed 26 was deposited in thin layers and is darker because of the additional organic material. Above these is bed 27, a thicker unit of lime muds.[1] Beyond bed 27, the Early Triassic rocks are a suite of shales and muddy limestones with occasional fossils. Sev-

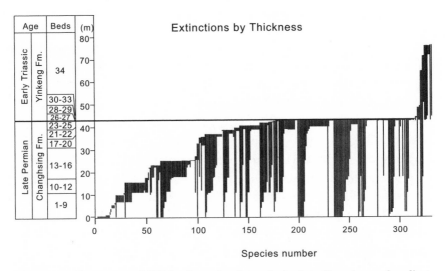

Figure 3.5 The ranges of 333 fossil species through the Late Permian and earliest Triassic at Meishan. Each line shows the beginning and end of species fossil ranges recorded at the five quarries. The column to the left shows a schematic of the sequence of rocks. Adapted from a figure in Jin et al. (2000).

eral meters above the boundary, rock surfaces are covered with the thin paper scallop known as *Claraia*, the characteristic fossil of the earliest Triassic.

Fossils are abundant within the latest Permian Changhsing Limestone, and over the past two decades a total of 333 species in 162 genera have been identified from fifteen different fossil groups, among them foraminifera and siliceous radiolarian micro-fossils, as well as rugose corals, lacy bryozoans, brachiopods, clams, snails, cephalopods, the tiny arthropods with a bivalved shell known as ostracods, some of the very last trilobites, tiny, teethlike conodonts, some fish, calcareous algae, and a few other groups. Some sixty-four different levels have been sampled across the five different quarry faces at Meishan, producing a highly detailed picture of life leading up to the Permo-Triassic extinction. Of the 333 species, 161 became extinct below the immediate boundary interval (figure 3.5). This might suggest that the extinction was a gradual one, even beginning in the early Changhsingian. But species disappear all the time, so we first need to remove the background rate of extinction and see if the number of disappearances at Meishan is higher than normal. In none of the beds below bed 24 does the rate of disappearances exceed 33%, which is within normal values. Between beds 24 and 26 almost all the remaining species disappear, an extinction rate of 94%.

On the face of it, figure 3.5 suggests a very rapid extinction in beds 24–26. But we have not considered the quality of the record. Paleontologists do not expect that the very last individual of a dying species will survive the lottery of preservation to make it into the fossil record. This means that the last recorded occurrence of a species will be earlier than the actual time at which the species became extinct. The more common a species, the shorter the gap is likely to be between the last fossil occurrence and the time the species breathed its last. For example, if we have two species, A and B, but we have recorded A from fifteen different levels in the Changhsingian Formation but B only twice, the last recorded position of species A should be closer to the point where it went extinct than the last record of species B.

This is just the sort of issue for which statistics is helpful, and my friend Charles Marshall from Harvard, expanding on previous studies by Peter Sadler at the University of California, Riverside, developed a suite of techniques to study this problem. This approach requires knowing not just the first and last occurrences of a species (the data shown in figure 3.5) but how many times a species has actually been recovered between the earliest and last occurrences. Without going into details, Charles's methods allow one to compute extensions to the range of a species so that there is a 50% or 95% probability that the true extinction point lies within the range extension. In other words, this statistical approach allows us to take the known occurrences of a species and correct the record to estimate where it was likely to have become extinct; we can then see if any of the corrected extinction points overlap with one another, or with other events, such as a volcanic ash bed, an impact horizon, or a change in the carbon cycle.

At Meishan, we decided that the generic data was more reliable and we chose to analyze this amalgamation of the species data. Wang Yue, then a young student of Jin Yugan's in Nanjing, did most of the actual work and developed a few extensions to Charles Marshall's original approach. She took the 265 species in ninety-three genera that had been found multiple times at Meishan, and then calculated the 50% and 95% confidence estimates. Then we tested how rapidly the extinction occurred, whether it was gradual, occurred in a series of steps, or was truly catastrophic. Using the radiometric dates from Meishan discussed in the next chapter, we were also able to convert the thickness of the rock to time, essentially rescaling the data so that we had the duration of species and genera in millions of years rather than thickness of rock, and to look at the rates in terms of absolute years. These results, published in the journal *Science* in 2000, showed a clear extinction peak in beds 25/26, occurring between 251.2 and 251.4 million years ago. (Changes in our estimate of the actual age of the boundary would shift this down to 251.6 and 251.4 million years ago.) The 50% confidence estimates for all ninety-three genera are consistent with a sudden extinction, but a more reasonable and equally well-supported conclusion would be that there was a sudden extinction at about 251.4 (now 251.6) million years ago, with a small number of species surviving but dying out over the following 1 million years.[2]

The abundance of marine rocks spanning the Permo-Triassic boundary also makes China an ideal venue for dissecting the fine-scale changes in sea level. Geologists have three different methods for unraveling the history of sea level: plotting changes in coverage of marine rocks across continents; studying individual localities where the sequence of shales, limestones, sandstones and other rocks reveal changing environments and depth of the water; and finally identifying packages of rock laid down at a single time, or depositional sequences. For example, as sea level rises, the seas will inundate formerly dry land and deposit a package of rocks—perhaps muds in the deep sea, limestones in near shore reefs, and sands near the shore. The surface along which the seas rose will become the lower boundary of a depositional sequence. Eventually the seas will retreat again, producing another surface that ends the sequence. This method of sequence stratigraphy relies upon identifying the surfaces that bound each depositional sequence. Summed across many localities in a region, and then across regions, geologists can distinguish regional changes driven by uplift or sedimentation from global changes in sea level produced by glaciations, changes in the rate of sea-floor spreading, or related processes.

Using the first approach of plotting changes in coverage of continents with marine rocks, the late Bill Holser from the University of Oregon, and his late collaborator Mordecai Magaritz of the Weitzman Institute in Israel (each of whom will figure prominently in chapter 7) toted up the types of Permian and Triassic rock for sixty-eight different marine basins around the world. Their results showed that on average, the oceans before the extinction had covered about 40% of the continents, but this coverage dropped to 8–13% near the Permo-Triassic boundary. Relying on fairly crude estimates of the slope of the continents (recall from chapter 2 that this is necessary to convert coverage to vertical distance), Holser and Magaritz estimated that sea level dropped about 280 meters at the boundary, one of the largest drops in the past 600 million years. Their data also suggested that sea level then climbed almost as rapidly through the Early Triassic.[3]

I faithfully noted this estimate in my 1993 book on the extinction, but gave far too little credence to work emerging from Paul Wignall at the University of Leeds, and Tony Hallam, at the University of Birmingham. They took the second approach to reconstructing the history of sea level, embarking on detailed fieldwork studying individual sections. They discovered that the estimates of Holser and Magaritz did not match the rocks at Permo-Triassic boundary sites. After carefully examining the sequence of deposits across Permo-Triassic boundary sites in different parts of the world, Paul and Tony challenged the orthodoxy of a drop in sea level.[4] Subsequent studies and full-scale analyses of the sequence architecture (the third approach) have been published over the past decade. Two Chinese groups, one led by Jin Yugan at Nanjing and the other by Professor Yin Hongfu at the China University of Geosciences at Wuhan, have used sequence stratigraphy to examine Chinese sea-level changes. Both groups established that sea level rose across the boundary, although they differed slightly on whether there may have been a slight drop in sea level before the extinction horizon.[5] There is now fairly conclusive evidence from both single localities and more regional studies that while sea level may have dropped in much of the world through the Late

Permian, this trend reversed about a half a million years or so before the beginning of the extinction. In China and in most tropical regions sea level was rising by the time of the extinction.

. . .

The village of Lalongdong lies only a couple hours north of Chongqing, one of China's largest cities. To reach Lalongdong, Jin Yugan and I took a taxi along the Jailing River, crossed a bridge to the east side, and turned onto a farm road that climbed slowly into the high mountains. The clouds and rain that had made fieldwork miserable for the previous couple of days slowly cleared, and after negotiating our way through an illegal roadblock established by local officials to raise revenue, we reached a small side road with a cluster of farms. Banana plants clustered around the small, neat houses. The climate is so mild in this rich agricultural region of Sichuan Province that farmers can grow three crops a year. It was mid-September and the farmers waved to us as we walked up through their fields of melons, beans, and squash toward the fossil Lalongdong reef. We have often had curious or perplexed receptions from the local people in other rural areas of China, but these farmers were accustomed to geologists stepping carefully through their fields.

The Lalongdong reef is one of the youngest Permian reefs anywhere in the world and although the geologic pilgrimages hardly rival Meishan or the Permian reefs of west Texas, give them time. The reef is now rotated up on its side, so the youngest part of the reef is to the left as one walks up the side of the hill. Jin and I took a path through the fields to the right, so that we could traverse to the left, meandering through bean fields as we walked through the reef from bottom to top. The reef was tens of meters thick, and as we studied the rocks with small hand lenses the fossils packed into the limestone were obvious. Fan Jiasong of the Institute of Geology in Beijing had carefully studied this mass of crinoid fragments, sponges, calcareous algae, brachiopods, and other fossils. The massive limestone continued to within about 1.5 meters (5 feet) of the Permo-Triassic boundary. An odd carbonate crust caps the reef, about 1 meter (3 feet) thick, with long, finger-

Figure 3.6 The carbonate crust atop the Lalongdong Reef outside of Chongqing, China. The Permo-Triassic boundary lies just above the hammer at the top of the massive limestone, which forms part of the unusual carbonates of the latest Permian.

like structures embedded in a clotted texture much like cottage cheese. The contact with both the underlying reef and the overlying Triassic shales is sharp, suggesting that whatever formed this crust developed quickly and then disappeared (figure 3.6). It was these last few feet and the mystery of what produced this clotted texture that brought us here.

In 2001 geologist Steve Kershaw, from the United Kingdom, and his colleagues reported finding the conodont *Hindeodus parvus* within the crust, confirming that the crust developed during the earliest Triassic, rather than the latest Permian (when Jin Yugan and I visited we thought the boundary lay above the crust).[6] Kershaw and his group also found the crust in other Permo-Triassic boundary sections in Sichuan Province. Similarly odd carbonate sediments have been reported from Iranian Permo-Triassic boundary sections as well. During the Late Permian, Abadeh, near the Hambast Mountains in central Iran, lay on a small microcontinent to the northeast of the Persian Platform, then a part of the Pangaean supercontinent, but far to the west of south China. A 1-meter-thick, inorganically precipitated deposit occurs just above the Permo-Triassic boundary and includes 10–20-centimeter-long crystals that were probably originally aragonite. Similar structures have also been reported from southern Japan.[7]

These carbonate cements were common in the Proterozoic a billion or more years ago, and well before the appearance of animals, when the absence of carbonate-secreting animals allowed carbonate in the oceans to build up to high levels. The carbonate came out of the water as an inorganic precipitate. Today such direct inorganic precipitation of carbonate from seawater is rare. Since animals with skeletons appeared in the Cambrian, carbonate levels have almost always been low enough to prevent inorganic precipitation. These crusts at the Permo-Triassic boundary may reflect a temporary imbalance in carbonate levels in the global ocean caused by the extinction itself. Or global warming could have reduced the solubility of carbonate in the ocean, leading to inorganic carbonate precipitation, as can happen today in warm tropical settings where evaporation is high, such as the Persian Gulf. Another alternative is that the precipitates were not inorganic at all, but formed by microbial activity, similar to the microbes that form stromatolites. Unfortunately in the 250 million years since the crusts formed, they have, like many carbonates, been recrystallized, a process that destroys the original structure. In this case the recrystallization has destroyed the fine structure that might reveal whether microbes aided formation of the crust. True stromatolites do overlie the crust at some localities in Sichuan, but this does not necessarily mean microbes formed the crust itself. Kershaw argues that the crust was formed by unusual ocean chemistry (high alkalinity produced by excess carbon dioxide) leading to a widespread carbonate precipitation event, although he remains agnostic about whether microbes were involved in its formation.

· · ·

Trilobites were the most abundant fossils of the Cambrian. They are also widespread in Cambrian marine rocks, and some groups of trilobites evolved fairly rapidly. These properties made them ideal for correlating Cambrian rocks between different regions.

Ammonites had the same properties in the Jurassic (named for the Jura Mountains in Europe) and form the basis of Jurassic fossil correlations. For other time periods, appropriate fossils were chosen. This process of correlation with fossils is known as biostratigraphy.

Recall our discussion of Roderick Murchison's recognition of the Permian in Russia. The top of Murchison's Permian was a sequence of nonmarine rocks. Although they contained fossils, nonmarine fossils are generally much harder to correlate to other parts of the world than marine fossils. (In fact, this nonmarine sequence remains difficult to correlate even today.) The lowest unit of von Alberti's Triassic was the Buntsandstein, a nonmarine sand that lacks fossils. Underneath the Buntsandstein lies a marine limestone known as the Zechstein, found across Europe and into England. In his original description of the Permian, Murchison recognized that the fossils within the Zechstein belonged to the Permian. Soon the Zechstein-Buntsandstein contact was taken as the Permo-Triassic boundary, and by the 1860s this had been roughly correlated in the Dolomite Alps of Italy with the contact between the Permian Bellerophon Formation and the Triassic Werfen Formation. The lowermost Werfen contains the paper-thin scallop *Claraia*. In G. L. Griesbach's studies of the geology of the Himalayas during the 1870s he found *Claraia* and with it the ammonoid *Otoceras* above rocks containing Permian fossils. Griesbach used *Claraia* to correlate the rocks of the Himalayas to the Werfen in Italy, and in time *Otoceras woodwardi* became the defining fossil of the earliest Triassic and thus the Permo-Triassic boundary.

As paleontologists continued to study the Permo-Triassic interval, other ammonoids, some Permian brachiopods, and a few other fossils proved their utility for correlation between regions on a single continent and between continents. Correlation depends on a turnover in distinctive, widespread fossils. The faster the group evolves, the shorter the time interval that can be defined using a specific fossil species. Permian ammonoids (figure 3.7), brachiopods, and clams evolve fairly rapidly, but by the 1970s a new group of fossils had emerged as contenders for finer biostratigraphic correlation: conodonts.

Figure 3.7 Ammonoids long served as the main biostratigraphic tool for correlating Permian and Triassic rocks within regions and between continents. This is the ammonoid Jin Yugan is photographing in figure 3.2.

Conodonts are tiny, jagged, toothlike fossils made of calcium phosphate and common in rocks from the Ordovician through the Triassic (figure 3.8). For decades paleontologists knew they must be part of some larger animal, but had no idea what that animal could be. In 1969 a geologist from Michigan State University unveiled a fossil at a scientific meeting showing conodonts clustered in the gut of the animal and claimed they were used as part of the digestive apparatus. At the conclusion of his talk, another paleontologist stood up and suggested this was a conodont-eating animal! In 1993 the true conodont animal was finally revealed as a primitive, soft-bodied relative of the vertebrates, but lacking a backbone, and the phosphatic conodonts turned out to be parts of the feeding structure in the mouth.[8]

Despite the long uncertainty over their true biological affinities, conodonts are ideal fossils for biostratigraphy. They evolved as rapidly as Cambrian trilobites or Jurassic ammonoids and were common in marine rocks around the world. During the Permian and Triassic most ammonoids were restricted to tropical waters, making correlations between the tropics and high latitudes very difficult. Conodonts did not suffer this problem. During the 1980s and

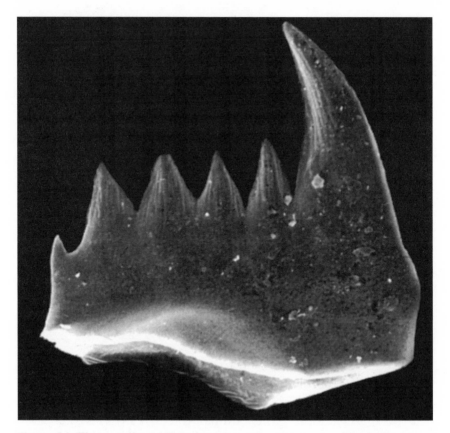

Figure 3.8 The conodont *Hindeodus parvus*. Conodonts are phosphatic micro-fossils that formed the chewing apparatus of small, eel-like animals. This spe-cies is now used to define the base of the Triassic, and the rapid evolution and wide distribution of conodonts have made them ideal biostratigraphic tools through much of the late Paleozoic and the Triassic. They became extinct dur-ing the end-Triassic mass extinction. Photograph courtesy of Heniz Kozur from Kozur (1996).

1990s a series of different biostratigraphic schemes using cono-donts were proposed for the Late Permian and Triassic. A cono-dont known as *Hindeodus parvus* (now *Isarsica parvus*) emerged as the standard fossil for the base of the Triassic, displacing *Otoceras woodwardi* from the honor.[9]

Over the past century and a half paleontologists have used am-monoids, bivalves, and now conodonts to progressively divide the Permian and Triassic into a sequence of series, stages, and, at the finest level of detail, biostratigraphic zones (figure 3.9). Although

Period	Series	Stage	Ammonoids	Conodonts	Fusulinids
Triassic		Griesbachian	Ophiceras Otoceras	Hindeodus parvus	
Permian	Lopingian	Changhsingian	Rotodisoceras - Pseudotirolites - Pleuronodoceras Pseudostephanites - Tapashanites Paratirolites - Shevyrevites Iranites-Phisonites	Clarkina changxingensis C. subcarinata	Palaeofusulina sinensis Palaeofusulina minima Gallowayinella meitienensis
Permian	Lopingian	Wuchiapingian	Sanyanhites Araxoceras - Konglingites	C. orientalis C. leveni	

Figure 3.9 The sequence of ammonoid, conodont, and fusulinid fossil zones used for biostratigraphic correlation around the Permo-Triassic boundary. Zones are named for the most characteristic fossil within them.

disputes continue about some of the correlations, the general framework has been accepted. Geologists have established standard reference sections for each of these intervals, with the Middle Permian reference a series of outcrops in western Texas and the Late Permian defined on sequences in South China, including Meishan.

· · ·

The Chinese work has greatly rejuvenated interest in the Permo-Triassic boundary, and this led to new searches for the Permo-Triassic boundary in other parts of the world. Following the principle of seek and ye shall find, new localities have been uncovered in many different areas (figure 3.10). In south China alone upward of fifty different localities have been studied by Chinese geologists, and dozens more are scattered around the world.

In addition to the use of fossils, geologists have developed new tools to correlate rocks. One of the most useful tools is based on the cycling of carbon through the oceans, atmosphere, and over

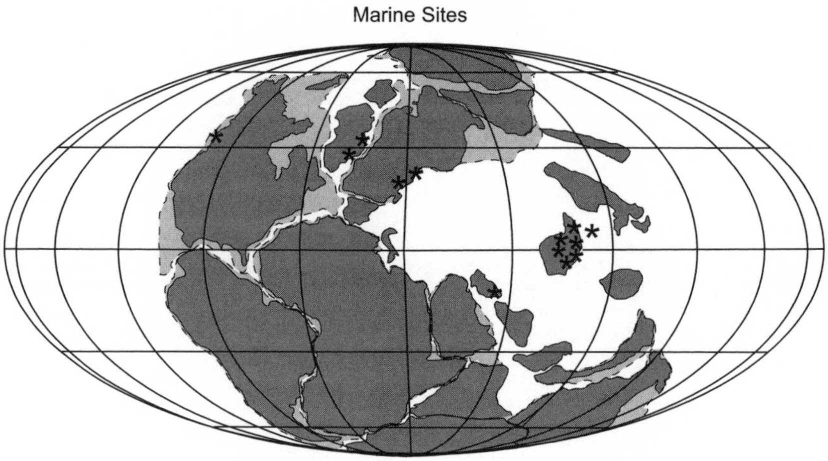

Marine Sites

Figure 3.10 Location of well-studied sequences of marine rock spanning the Permo-Triassic boundary in different parts of the world, on a map depicting the configuration of the continents in Late Permian time. There are over fifty sections in south China alone, so only a few are shown here.

a longer timescale, through rocks as coal, oil, limestone (made up of calcium carbonate). Mass extinctions and other biotic crises are often associated with dramatic swings in the carbon cycle, and these are preserved in the rock record. We will discuss this in more detail in chapter 7, but these abrupt changes in the carbon cycle serve as a distinctive marker of the boundary. Once geologists have zeroed in on a sequence of rocks that appears likely to contain the boundary, collection and analysis of a series of rock samples should reveal the distinctive chemical fingerprint as the isotopic ratio drops. As we will see in chapter 7, areas in higher latitudes, including New Zealand and Antarctica, appear to preserve multiple negative carbon shifts, and there are a series of dramatic swings in the carbon cycle preserved in Early Triassic rocks. So in some cases, the carbon isotopic record must be used in conjunction with other evidence.

Volcanic ash beds at the Permo-Triassic boundary at Meishan opened a whole new chapter in research on the causes of the extinction by providing us with the first reliable view of the rapidity of the extinction.

CHAPTER 4
It's a Matter of Time

Sam Bowring was sitting in the California sunshine early one morning in 1995, looking uncomfortable in a coat and tie. Despite his stature as a professor at MIT, Sam remains an unrepentant field geologist, with a bushy beard, nervous energy, and a 1970s rebellious streak.

"So, there are all these ash beds in the Permo-Triassic sections in South China—are you interested?" It wasn't my best opening line, but I was pretty sure it would work.

"What do you mean *all these ash beds?*"

Sam and I had met the day before, but for several years I had known of his exceptional work dating volcanic ash beds using the decay of uranium to lead in volcanic ash beds. Sam is one of the best geochronologists, as age-daters are known, in the world. His lab at MIT has been a pioneer and leader in pushing analytical precision. We were attending a meeting held by the U.S. National Academy of Sciences bringing together young scientists from very different fields for several days of scientific presentations and discussions. Sam and I had each given talks the previous day on the origin of animals during the late Neoproterozoic-Cambrian period. Sam had discussed his lab's recent work providing new dates for this interval by very precise analyses of 544 and 532-million-year-old ash beds from Siberia. These new dates had established the first temporal framework for the initial explosion of animals. I hoped he could do the same for the end-Permian mass extinction.

I later learned that "No dates, no rates" was Sam's motto, and that fit perfectly with my interests in geochronology. Paul Renne of the Berkeley Geochronology Laboratory had already dated the volcanic ash bed close to the Permo-Triassic boundary at Meishan (see chapter 3) at 251 million years ago, plus or minus a bit over 3 million years. While it was great to know the age of the boundary, Sam and I knew that by dating the many other ash beds preserved above and below the boundary, we had a good chance of establishing how rapidly the extinction occurred. On that sunny morning we quickly hatched a plan to collaborate with Jin Yugan, and by 1998, we announced that the extinction had happened in less than 500,000 years, and perhaps even more rapidly. Resolving time to that level of precision was beyond the imagination of the geologists who first established the Permian. Extending our discovery at Meishan beyond south China to other regions depended on our ability to correlate rocks of the same age in disparate parts of the world.

· · ·

Fossils alone allow geologists to sequence the order of rocks, but this produces only a relative time scale: A is older than B. Identifying absolute ages, or the age in years when an event occurred, had to await the discovery in 1896 that uranium was radioactive. Geologists soon realized that radioactive minerals provided a means to establish the age of geologic events, from the creation of mountains to volcanic eruptions. This discovery sparked an ongoing effort to refine the geologic time scale.

The rocks at Meishan represent the ultimate answer to the geochronologist's prayer, although I hardly realized this when I told Sam about the many ash beds at Meishan. Since 1995 Sam and I have spent a great deal of time wandering around the world searching for ashes (on "Sam and Doug's Excellent Adventures"). Often we cannot even find an ash, or when we visit spots where other geologists have reported ashes, the ashes do not contain the minerals needed for radiometric dating. Or, through the perversity of the geological gods, we find a beautiful ash, but in a sequence of rocks so devoid of fossils that we have no idea how to connect a radiometric age to established biostratigraphic zones.

At Meishan, everything comes together in one beautiful package: An ash demarcates the end-Permian mass extinction, and as we saw in the previous chapter the fossils have been thoroughly described. The rocks preserve an unaltered record of the changes in ocean chemistry through the extinction and the early phase of the subsequent biotic recovery, and then there are "all these ash beds." The quarry faces at Meishan preserve at least four ashes below the boundary, the boundary ash, and then at least four more in the Early Triassic. Some of Jin's students at Nanjing have found that many more, thinner ashes may be present as well. By dating each of the ash beds, Sam can check the reliability of his results. If all goes well, the dates should come out in order, with the oldest ash on the bottom. If the ages do not come out in the right order, he knows that something has gone wrong. Even more importantly, since some ashes formed before the extinction, with dates in hand we can establish the rapidity of the extinction.

Carbon-14 radiometric dating is the best-known dating method among the general public from its use in dating recent archeological discoveries. But it is useless beyond 50,000 years or so. Geologists looking deeper into the past concentrate on other radiometric dating systems. All radiometric dating depends on the spontaneous decay of the isotopes of a parent element into a different, daughter element. (The isotopes of an element have the same number of protons in the nucleus but differ in the number of neutrons; carbon-12 has six protons and six neutrons while carbon-14, for example, has two more neutrons than carbon-12.) Uranium-238 and uranium-235 are among the elements that undergo radioactive decay through nuclear fission, releasing protons in the process, so uranium-238 and uranium-235 eventually become lead-206 and lead-207 respectively. Potassium-40 decays to argon-40 via a process known as beta decay, as a neutron converts to a proton and an electron is ejected, as in the conversion of carbon-14 to nitrogen-14, and is also useful in radiometric dating.

Each of these systems has a characteristic rate of decay that can be measured in the laboratory. From the rate of decay physicists and geochronologists can calculate the half-life, the length of time required for half of the original number of atoms to decay to the daughter product. Knowing this rate and the amount of the parent and daughter elements in a sample, geochronologists can calculate the age of a rock or mineral. The decay constants are usually accurately known although as geochronologists resolve smaller and smaller increments of time, the decay constants are continually under scrutiny. They do not change with time as some creationists have argued but our ability to measure the rate of decay gets better with time. However as I write this book there appears to be a consistent bias of a little less than 1% between rocks dated by K-Ar and those dated by U-Pb zircon with zircon falling on the older side. While this may not seem like a lot, for rocks 250 million years old it is 2.5 million years or eight to ten times the precision possible with U-Pb zircon geochronology. So for now we must think of Ar years and U-Pb years when comparing ages.

The critical issues for determining the absolute age of a rock are finding a rock to date, and then measuring the amount of parent and daughter isotopes in it or a constituent mineral. It turns out that rocks that have been exposed at the surface of the earth for very long are often modified by interaction with water. In this case many of the minerals are altered so one must select a mineral that resists alteration.

Volcanic ashes from magmas rich in silica generally contain the mineral zircon. Zircon sands are mined around much of the world, attesting to the toughness of the grains. Tiny needles of zircon are often only 100–200 microns (a tenth of a millimeter) long (figure 4.1). When they form they contain small amounts of uranium, but no lead since the lead atoms are excluded. Over time, the uranium in the zircon crystal decays to lead and thus the crystal can be thought of as a time capsule. If we measure the amount of U and Pb present today and we know the rate of decay of the uranium isotopes, we can calculate a date for the crystal.

The accuracy of radiometric age determination requires that the only change in the parent/daughter ratios of uranium and lead in a zircon be due to radioactive decay, what geochronologists

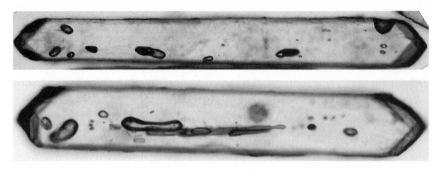

Figure 4.1 Zircon grains similar to those recovered from Meishan by Sam Bowring's lab. Photograph courtesy of Sam Bowring.

call closed system behavior. If lead or uranium is either lost or gained the date will be erroneous. The two different radioactive isotopes of uranium (U-238 and U-235) have different rates of decay (the half-life for U-238 is about 4.5 billion years and for U-235 is about 750 million years). Comparing the results from each decay series provides an internal check on whether the system was closed. If the zircon has remained as a closed system, the two calculated dates should be approximately equal. Much of Sam's proselytizing for uranium-lead dating versus other methods stems from the added reliability of this internal check and the resistant nature of zircon grains.

The actual measurements of isotopes of uranium and lead are made on a mass spectrometer whose inner workings are best left to Sam and his colleagues (he did let me touch his machine once—I think I was supposed to feel honored). A zircon analysis involves a lot of painstaking work that starts with the separation of zircon crystals from volcanic ash followed by careful selection of the clearest and best grains and finally their dissolution in a pressurized vessel at temperatures of more than 200°C for several days. Then each dissolved crystal is treated so that the U and Pb can be separated from all the other elements. Since many zircon crystals from Meishan contain as little as 10–20 picograms (10^{-12} gm) of lead it is crucial that no lead is added in the process from laboratory contamination. Lead is in everything: water, your eyelashes, hairs, and food. So the laboratory process requires careful handling in a very clean environment. In most state-of-the-art labs it is possible to add less than 1 picogram of lead to the analysis.

Our first trip to China was in 1996. In early 1997 Sam called me with the first results: "1.2 billion years. That's the age of the boundary." Sam knew perfectly well the age was way off. They had isolated some beautiful zircon crystals but by coincidence the first ones they analyzed were inherited from much older rocks. When Sam studied the regional geology, he realized that the Permian magma must have come up through very old rocks formed during the origin of the continental crust in southern China. Somehow older zircons were incorporated in the volcanic eruption that produced the erupted ash bed. Sam and his team quickly isolated other zircons that produced a more reasonable age. Until recently mass spectrometers were not sensitive enough for analyses of individual crystals and geochronologists were forced to study a population at a single time. The preservation and incorporation of older zircons is one reason most analyses now use single zircon crystals. Many problems of inheritance are far subtler, requiring geochronologists to study each zircon individually.

My favorite memory of our 1996 trip to Meishan came near the final day. Sam, Jin, Sam's postdoctoral fellow Drew Coleman, and I had collected ashes from six different levels, with several replicates of some of the ashes (figure 4.2). The Permian ashes were wet and gooey, like good clay. We began by clearing off the overlying rock with a geological pick (christened the Peacemaker for some reason) and uncovering a nice layer of ash. We then used a big putty knife to carve the ash into blocks and wrap them in plastic bags. Another of Sam's many sayings is that you can never have too much ash. At the time he had no idea how abundant zircons would be within the ashes, and thus how much material we would need. That day Sam and I sat on the boundary looking at an ash in bed 27b. We had many samples of the boundary beds (25 and 26) and good samples of bed 28, about 17 centimeters above the boundary. We stared at the ash in 27b, and Sam, in words that have come to haunt him (because I have not let him forget them), said "Nah, there is no point in collecting it, we'll never be able to get high enough quality dates to tell this apart from the other ashes."

Oh ye of little faith. After months of incredibly hard work, Sam and his group produced dates on the Meishan ashes with uncer-

Figure 4.2 Photograph of our group collecting ash beds at Meishan. The two limestone cliffs in the center of the photograph are the last rocks of the Permian. Zhu Zhuli, to the left, has his feet on the Permo-Triassic boundary. Sam Bowring is above, standing on one Early Triassic ash bed and pointing to another.

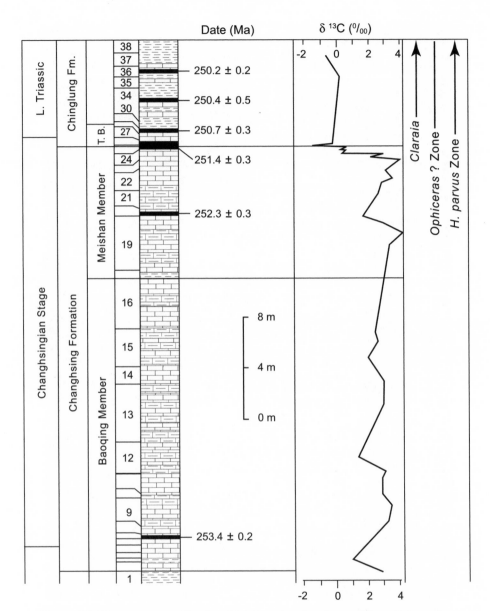

Figure 4.3 The sequence of beds at Meishan showing the position of the major ash beds and the original radiometric dates from Sam Bowring's lab as reported in the 1998 paper. We now think the best estimate of the boundary ash is 251.6 million years ago.

tainties of as little as 200,000 years (figure 4.3). When we published the radiometric dates from South China in *Science* in 1998, we had a good date from bed 20, about 4.3 meters below beds 25/26, at 252.3 +/– 0.3 million years ago, and the boundary to 251.4 +/– 0.3, a difference of 900,000 years. But since fossils continued to be diverse well after bed 20, the extinction must have happened between these two beds, and largely within the last 50 centimeters. Based on estimates of how rapidly sediment was deposited at Meishan, we argued the extinction occurred in much less than 1 million years and possibly in less than 160,000 years (based on estimates of sedimentation rates between the dated ash beds). The lowest ash we collected, from bed 7, corresponded with the base of the Changhsingian Stage, the last stage of the Permian. There had been no good estimates for the length of the Changhisingian, but we showed that it lasted about 2 million years.[1]

The precision of Sam's results staggered me. During my time in graduate school we figured we were happy to know the age of Paleozoic fossils to within 5 million years. And even that often involved a fair bit of guesswork. Sam's results compounded an error I made during our fieldwork. I was so wedded to the idea that the mass extinction happened over millions of years, that it was not until we were sitting at Meishan (my second visit actually, since I had spent a rainy afternoon there with Jin in 1994) that Jin convinced me that fossils remained abundant throughout the Changhsingian, until right below the boundary. Fortunately the two decades of study of the Meishan by Jin's group at Nanjing, and another headed by Yin Hongfu thoroughly documented the distribution of fossils. So the critical point is that our radiometric dating provided *maximum* estimates of the amount of time involved in the extinction. The extinction could have occurred in much less than 160,000 years, but we could not reliably estimate how much less.

We further refined our estimates of the duration of the extinction by looking at the change in carbon isotopes. As detailed in chapter 7, the shifting balance in the ratio of the two carbon isotopes that do not decay over time, C-12 and C-13, chronicles shifts in the carbon reservoirs. The sharp anomaly found at many other Permo-Triassic boundary sites around the world coincides with beds 25 and 26 at Meishan. This increases our confidence that the ages obtained in Meishan apply elsewhere. The dates also tell us the rapidity of the carbon shift. In 1987 a core was drilled through a Permo-Triassic boundary section in western Austria. Based on the number of conodont zones over which the shift occurred and a guess about the duration of each conodont zone, the geologists studying the core estimated that the isotopic shift occurred over about 3 million years. With the radiometric dates from Meishan we showed that the isotopic shift happened in less than 160,000 years, and perhaps far less. Our study was the first of its kind in terms of scope and number of analyses.

As is usually the case, others wanted to see how well they could do on the same rocks. Most of the analyses involved analyzing very small amounts of lead and each was a challenge. In particular, to minimize the possibility of open system behavior, most zircon geochronologists have found that by removing the outer part of the zircon crystal by mechanical abrasion, the result is seemingly higher precision analyses. Roland Mundil and his colleagues at the Berkley Geochronology Institute (where Paul Renne also works) have dated some of the same ash beds from Meishan and suggested the age for the boundary was closer to 253, 254, or even 255 million years ago. In addition, the ages they calculated for each bed appeared to increase up the sections, so the stratigraphically higher beds had older ages, the opposite of expectations. Their main thesis was that some of the grains had experienced open system behavior and lost lead, so Mundil interpreted the oldest ages as the best approximation of the age. However, it is not uncommon for ash beds to contain zircons that are slightly older than the eruption age. Thus it requires many analyses for each bed to unravel all the possible complexities.[2]

We had another check on the reliability of our results from Meishan. During our 1996 fieldwork in China we also visited several Permo-Triassic boundary sites in Guangxi Province, in southernmost China just north of Vietnam. Lao Jin had sent some of his students on a long swing through the area before Sam and I arrived, checking reports of volcanic ash beds from old geological reports. One of the most memorable spots was a railroad embankment where I heard the distant whistle of a train as we collected an ash.

Figure 4.4 The thick volcanic Permo-Triassic boundary ash at Metan, in Guang-xi Province, China, with a geologic hammer for scale. The ash is actually three separate units, from the thick bed near the bottom of the picture up through the layer behind the hammer. The Permo-Triassic boundary is placed just above the top layer.

Moving off the track, I expected a diesel locomotive. Just in time, I realized that diesel trains do not produce huge billowing clouds of white smoke, or sound steam whistles. To the laughter of the Chinese students, I grabbed my camera and ran along the wooden trestle taking pictures as the train thundered by. Coal is so abundant in southern China that steam trains remain economic, and that locomotive must have been hauling freight since the 1930s.

Some of this coal comes from two other collecting spots we visited on that trip, at Matan and Panglaitan along the Hong Sui (Red) River, the river that flows into Guangzhou (formerly Canton). The river rises tens of meters during the typhoon season, scouring the outcrops clean, and then falls during the winter dry season, so the rocks are beautifully exposed. The rocks dip gently to the north, revealing a virtually compete section from the Middle Permian well into the Triassic. At Matan we found a spectacular series of ash beds just below the Permo-Triassic boundary (figure 4.4). Within each ash the largest, heavier material is found at the bottom, and progressively finer material is found toward the top. Such fining upward sequences are characteristic of material that has fallen out from an ash cloud, and convinced us the ash bed had not been moved by wind or water after it was first deposited. Amazed at our good luck, Sam and I collected four samples here, with the top two essentially identical to the boundary ash at Meishan. When Sam's group finished the analyses the results came out at 251.6 +/- 0.1 Ma, 251.7 +/- 0.2 Ma, and 251.6 +/- 0.1 Ma (the lowest two samples were from the same level and were combined into the first number in the 1998 paper). Spot on the results from Meishan, and because the uncertainties are less, these are the best estimates for the age of the Permo-Triassic boundary. The results from Matan were essentially identical to those from Meishan. Reaching the same conclusions from ash beds over 1,000 kilometers apart provides further confidence that our initial results were reliable.

More recently Jim Mattinson from the University of California at Santa Barbara, (and a member of my thesis committee in graduate school) has developed a technique that seems to eliminate the parts of the zircon crystals that have suffered open system behavior. Jim has been a pioneer in uranium lead geochronology for more than thirty years. While he was at the Carnegie Institution in Washington in the early 1970s, he developed a method to produce the acids needed for zircon analysis with very low levels of uranium and lead. This revolutionized laboratory analysis by allowing study of the very small amounts of uranium and lead in single grains of zircon. Well more than thirty years later Jim has once again revolutionized the field by developing a technique that heats the zircons at 800°C for two days followed by leaching in hydrofluoric acid. This leaves the closed system parts of the zircon grains behind and removes the more problematic bits, producing analyses of higher precision and accuracy, another major advance.

Mundil and his group recently published a paper on the Permo-Triassic boundary at a locality in Sichuan province called Shang-Shi where the paleontology is not as well constrained as at Meishan.[3] They used Mattinson's technique to produce an age for the boundary that is approximately 252.6 Ma, or about 1 Ma older than our results for Meishan. Their conclusions about the duration of the extinction are identical to ours but are shifted 1 million years older. Sam and his group now think that many of the initial dates from Meishan are too young but this simply shifts the absolute ages without affecting the conclusion about the duration of the extinction.

Interlaboratory comparisons are becoming crucial, something geochronologists never had to worry about until recently. When two labs produce different dates for the same rock, there are a number of possible explanations. It could be that the ash beds of Meishan have open system behavior and the dates are too young. In addition, there could be differences between labs in how they have calibrated their technique. We are coming to appreciate that these interlaboratory differences are significant and that the geochronological communities must work together to resolve them.

· · ·

Other geologists have found our results from Meishan too coarse, rightly noting that testing claims for an extraterrestrial impact at the boundary really requires dividing time on an even finer scale. Michael Rampino of New York University has long been an advocate of impact, and in 2000, he and his colleagues decided to use the Milankovich cycles of the Earth's orbit, as preserved in rocks across the Permo-Triassic boundary in the Alps, to provide even finer scale-time resolution. Milankovich cycles are well-known wobbles in the tilt of the Earth's axis and the eccentricity of the Earth's orbit around the sun. Today these cycles are 100,000; 41,000; and 21,000 years, and they governed the waxing and waning of ice ages over the past 2 million years. Geologists claim to have found at least some of the cycles recorded in rocks well back into the Paleozoic, although how far back these periods hold remains controversial. Critical to testing claims of Milankovich, cyclicity is having sufficient high-resolution radiometric dates to establish which of the three cycles one has found in the rock record, establishing that the cycles are truly Milankovich cycles. Apparent cycles are ubiquitous in the rock record, but not all that glitters is gold.

In the Permo-Triassic rocks from the Alps, Rampino and his colleagues counted the number of cycles and claimed they show that the extinction happened in less than 60,000 years, and perhaps less than 8,000 years.[4] The cycles are hard to see in the data (a colleague has noted that they are impossible to see in the rocks in the field), although the statistical analysis seems to bear out their claims. I think the most significant problem is determining which of the various cycles, if any, they have identified. As much as I would like to be able to refine time down to tens of thousands of years, the data does not appear to be sufficient to confirm the claims.

The next challenge for Sam and his lab is to see just how precisely one can determine the ages of these ash beds. In the decade since Sam and I started collaborating on this problem, new techniques have made it possible to analyze ever-smaller amounts of zircon with ever-greater precision. The challenge now is whether this can be done at the ± 100,000-year level. If geochronologists could split time this finely it might be possible to evaluate Milankovitch cyclicity at a much higher level of accuracy.

· · ·

With the age of the Permo-Triassic boundary in hand, radiometric dating can also be useful in establishing whether other geological events occurred at the same time. In 1992 came the announcement that the Siberian flood basalts erupted at the same time as the end-Permian mass extinction. A joint American-Russian group reported radiometric dates of 248 million years ago plus or minus about 4 million years on a volcanic intrusion cutting through the flood basalt in the Noril'sk region. Since the flood basalt must have erupted before the intrusion that cuts through it, the age of the intrusion is younger than the flood basalt, thus providing a minimum age estimate for the flood basalt (figure 4.5).

The onset of the Siberian volcanism had been dated in 1991 by Paul Renne to 248.4 million years ago plus or minus 0.3 million years (in argon years), and thus the overlap between the Meishan and Siberian dates was too strong to ignore. Although Renne's result seems like a more precise date, there were some problems.

Figure 4.5 Field photograph of the Siberian flood basalts. Photograph by Mark Wilson.

Radiometric dating requires a precise knowledge of the rate of radioactive decay. If this decay constant is off even slightly, the results will be off as well. Periodically geochronologists and physicists refine their estimates of the decay constant, which requires recalibrating earlier results. In 1995 Renne recalibrated his 1991 results on the Siberian flood basalt to 250.0 plus or minus 1.6 million years: an older age with a greater uncertainty. In the 1995 paper, Renne and some Chinese geologists also revealed a new analysis of the volcanic ash at the Permo-Triassic boundary: 251.2 million years ago plus or minus 3.4 million years. With the recalibration of the ages for the Siberian flood basalts, the dates for the eruption of the flood basalt and the Permo-Triassic mass extinction in China coincided. Since 1995, several additional studies of the Siberian flood basalts have been published.[5]

Most efforts to determine the dates and duration of the eruption have focused on the Noril'sk area where active copper-nickel-sulfide mines have established a good geologic framework for further work. An intrusion through the lower third of the volcanic pile has been dated at 251.2 million years (plus or minus 0.3 million years) in argon years. To the northeast, the Maymecha-Kotuy region includes a sequence of volcanics that overlap with those from Noril'sk as well as some younger rocks with a total thickness of 6,500 meters. Sandra Kamo of the Royal Ontario Museum used the single-crystal uranium-lead method to provide the most accurate information on the duration of the eruptions. So here we switch from argon years to uranium lead years. The lowermost volcanics of the Arydzhansky Suite date to 251.7 million years ago plus or minus 0.5 million years. Rocks about 2,000 meters higher in the sequence, part of the Dekansky Suite, formed as a single 1,400-meter-thick volcanic flow during the waning stage of the entire volcanic sequence and yielded a date of 251.1 million years ago plus or minus 0.4 million years.[6] The relationships between different volcanic units at the top of the volcanic sequence in the Maymecha-Kotuy area are complex, and there may be another 1,000 meters or more of volcanic rocks above the Dekansky Suite, so the eruptions may have extended after 251.1 million years ago. Setting aside the uncertainties in the analyses for the moment, it appears from the work of Kamo that this massive sequence was deposited in about 600,000 years, at just the same time as the Permo-Triassic boundary in south China.

We have not heard the last of this controversy. In their recent paper on the Shangsi-boundary locality, Mundil ignored the work of Kamo and declared that the Siberian Traps and the extinction were coincident but they did it by correcting their unpublished argon data from Shangsi by 1% to make them 252.6 Ma. We will see if this is correct but it again raises the issue of interlaboratory comparisons. Ten years ago no one would have thought we would be arguing about a few hundred thousand years or that we would have to talk about time in argon and uranium-lead years! But this means that our internal analytical precision is so good that small differences in calibration between labs are now much larger than error bars for a single date.

Even without radiometric dating there is other evidence that Siberian volcanics formed relatively quickly. If there had been considerable time between eruptions of the various flows, geologists would expect to find evidence of erosion, the development of soils, or other features. While soils are found between some eruptive flows, the soils appear to have been relatively short-lived. The record of the Earth's changing magnetic field provides another, independent means to assess how rapidly the flood basalt formed. During revolution in plate tectonics geologists discovered that the earth's magnetic field frequently reverses direction. The magnetic field is driven by circulation in the Earth's molten outer core. Periodically this magnetic dynamo collapses and then reforms, flipping the direction of the magnetic pole. A compass would point to the south magnetic pole rather than the north after such a reversal. Sensitive analyses can detect this pattern of fluctuating magnetic direction, and volcanics are a particularly useful record of the changing magnetic field. Over 4,000 samples have been collected through the 3,500-meter-thick Noril'sk volcanics, and all samples have the same magnetic polarity. Without radiometric dates geologists cannot directly determine how long a single magnetic polarity interval lasted, but most intervals of constant magnetic polarity in younger rocks last a few hundred thousand to a million years. In the case of the Siberian eruptions the entire sequence of lavas was estimated to have formed in about 600,000 years.[7]

Near Kunming City in Yunnan Province, China, is the Emeishan Large Igneous Province, an earlier Permian flood basalt. Today the flood basalts cover an area of some 500,000 square kilometers with thickness of a few hundred meters up to 5 kilometers. Heavy erosion in this tropical area makes it difficult to reconstruct the original extent of the volcanism, but it seems far smaller than the Siberian flood basalts. Chinese geologists have known for years that the Xuanwei Formation, a Late Permian unit, overlies the Emeishan volcanics, and thus that the volcanism occurred before the end-Permian mass extinction. But how much before? A group of Hong Kong geologists recently dated two igneous intrusions associated with the Emeishan at 258.7 million years ago plus or minus 1.5 million years, and 256.0 million years ago plus or minus 1.0 million years. Their suggestion that this corresponds to the earlier, Middle Permian phase of the twin mass extinctions might be right, but it requires a bit of divination since we do not know the age of the extinction. I suspect this correlation will not hold up when radiometric ages are available for the Capitanian-Wuchaipingian boundary. It is nothing but a guess, but based on the geology, I would estimate the Capitanian to be only 3–5 million years long. This would leave the Emeishan LIP in the early Wuchiapingian.[8]

Some sediment also contains iron minerals and preserves a record of magnetic reversals. The sequence of magnetic reversals has been widely used to correlate rocks in different parts of the world. A longer sequence of magnetic reversals produces sort of a bar code: north-south-north-south, of varying durations. There are two long intervals during the past 600 million years when almost no reversals occurred. One of these is known as the Permo-Carboniferous reversed polarity superchron (a chron being a single interval of magnetic polarity, and a superchron a longer interval of stable polarity). This long magnetic quiet zone ended during the Middle Permian, and was followed by an interval of frequent reversals, lasting into the Triassic. There may be as many as thirteen magnetic reversals through the Late Permian. Such a rich record is potentially a wonderful tool for correlating different sections around the world. In this case there is almost an embarrassment of riches, for there are so many reversals that fossils, dateable volcanic ash beds, or other information is needed to distinguish one reversal from the next.

During the Changhsingian a compass would point toward the North Pole, a period of normal polarity, then reverse and point south during a very short reversed interval just before the Permo-Triassic boundary, then point north again during most of the Greisbachian stage of the Early Triassic. This pattern can be very useful in evaluating the continuity of sedimentation at a single locality: the short reversal will only show up if sedimentation was fairly continuous. Many sequences in South China, the southern Alps, and elsewhere have passed this test, increasing their value for establishing the events leading up to the extinction.[9]

The geologic time scale is always a work in progress. The quality of the time scale improves as new geochronological techniques are developed, new material to date is discovered, refinements are made to biostratigraphic correlations between different regions, and advances are made in correlation methods. The latest Permian is fairly well constrained with radiometric dates and we think we know the age of the Carboniferous-Permian boundary, but the middle of the Permian still has few good dates (figure 4.6). Figure 4.7 shows the progressive refinement of the Permian and Triassic

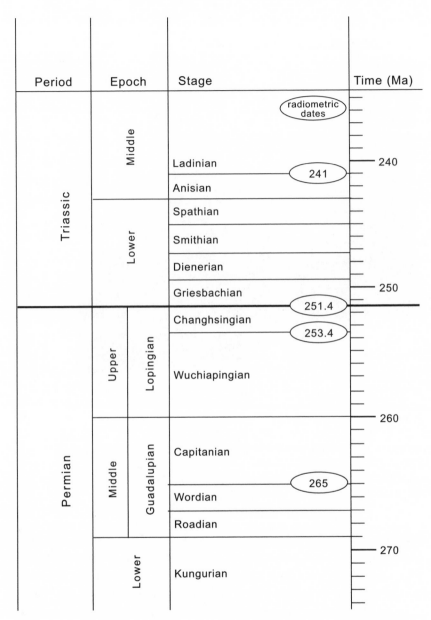

Figure 4.6 Timescale for the Permian and Triassic, showing the available radiometric dates.

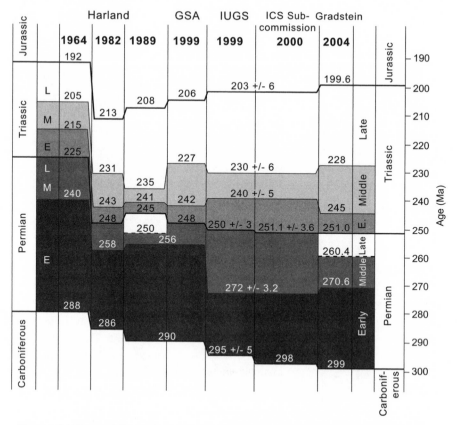

Figure 4.7 Progressive refinement of the Permian and Triassic portion of the geologic timescale from 1964 to 2004. Harland refers to the various editions of the Cambridge Geological Time Scale (Harland et al., 1964, 1982, 1989); GSA is the 1999 Geological Society of America timescale; IUGS and ICS are different versions of Remane (2000); and Gradstein is Gradstein et al. (2004).

timescale over the past century. Many of the early dates were interpolated between measured dates (which were often pretty poor), and interpolation is often just a fancy scientific term for a guess. What is striking to me, is how much there is left to learn in 2004.

The picture is not much different for many other intervals of geologic time. In younger rocks, say the past 100 million years, the timescale is fairly well constrained with radiometric dates and many different correlation techniques. This is partly because there is much more young rock available to geologists, and because the ocean basins provide long, virtually continuous record of changing magnetic patterns since the Cretaceous period. The success of our efforts to produce a high-resolution geologic timescale for the end-Permian mass extinction has encouraged Sam and me to extend our efforts to other critical events in Earth history. We have started a project dubbed EARTHTIME to produce a highly calibrated timescale back to 600 or more million years ago. The technology to accomplish this is now at hand, and we believe that such a cooperative project, encompassing paleontologists, geochronologists, stratigraphers, and many other geologists can achieve this goal by 2015.

The generation of so many new dates from Meishan, Matan, and other localities I have not mentioned has provided paleontologists their first real temporal framework for exploring the end-Permian mass extinction. I am sure that many surprises await us, but we can finally organize events not just in relative order, but also place them in absolute time. As with our efforts at the Permo-Triassic boundary, the objective is not simply to generate a higher resolution timescale. Instead, it is only by extending the resolution we have at the boundary to the rest of the timescale that we can examine a host of other events, establish their rates, and so understand the processes that caused them. As we will see in the next chapter where we discuss the patterns of extinction among animals in the sea, we are finally in a position to determine how rapidly the extinction occurred. In doing so we will also be able to toss out a few more of the extinction hypotheses. No dates, no rates.

CHAPTER 5
Filter Feeding Fails

Changing sea level, bursts of carbon dioxide or methane saturated water, the collision of an extraterrestrial object, massive volcanism. Each suggests different patterns of extinction. Which groups of animals suffered the greatest extinction may provide important clues as to the cause of the extinction. Did organisms in shallow water experience greater extinction than those in deeper waters? Then the trigger for the extinction must have had a larger effect in shallow waters. Were mobile species and burrowers more likely to survive than species attached to the sea bottom? Did predators and other species at the top of the food chain suffer greater extinction than those at the bottom? Can we detect any geographic pattern to the extinction, with greater extinction in Europe, China, North America, or in the tropics?

The end-Permian extinction imposed a major discontinuity on the history of marine life, eliminating the dominant clades of the Paleozoic, and allowed some of the surviving groups to establish the new ecosystems of the post-Paleozoic. John Phillips, the English paleontologist we met in chapter 2 first recognized the distinctions between Paleozoic and post-Paleozoic faunas in 1840, but Jack Sepkoski placed this on a more scientific basis. During the 1980s Jack used his compendium of the first and last occurrences of marine families to analyze the diversity history of the past 600 million years. I briefly discussed this in chapter 2, and figure 2.3 shows the resulting diversity pattern. Jack also used statistical techniques to identify three groups of organisms, each of which dominated marine ecosystems during different intervals of time, which he named evolutionary faunas (figure 5.1). Working backward in time, the Modern Evolutionary Fauna of clams, gastropods, crabs, bony fish, sharks, and a few other groups first appeared in the Ordovician, but accounted for a fairly small part of Paleozoic diversity. The Paleozoic Evolutionary Fauna of articulate brachiopods, corals, ammonoids, crinoids, and the lacy bryozoans is far less familiar to us because they suffered such catastrophic extinction at the end of the Permian. Some 72% of these families vanish but only 27% of families belonging to the Modern Evolutionary Fauna disappear. The Cambrian Evolutionary Fauna, the third in the series, encompasses the trilobites, odd echinoderms, molluscs, and other animals of the Cambrian. The last few trilobites made it to the Permian before giving up, but the Cambrian Fauna is fairly insignificant for our purposes.[1]

This dramatic disjunction between groups that survived the end-Permian extinction and those that either disappeared or became insignificant demands explanation. In this chapter I will focus on the members of the Paleozoic and Modern faunas, their fate during the Permian, and on three different ways of separating winners and losers of the extinction: ways of making a living, latitude, particularly the possibility of greater extinction on reefs; and physiology. Most of the major groups in the Paleozoic Evolutionary Fauna lived attached to the sea bottom and filtered small plankton and organic material out of the water. Paleontologists usually call these forms epifaunal (living on the sea bottom), sessile (attached) filter feeders. Modern groups are more mobile, and have vastly more predators. Were Paleozoic groups targeted because of their ecology, or was some other factor involved? Many paleontologists have often claimed that reefs are particularly susceptible to extinction, and if epifaunal filter feeders were more common on reefs, and reefs were particularly impacted, then this might account for the pattern of extinction, not filter feeding. Finally, two of my colleagues have recently suggested that the

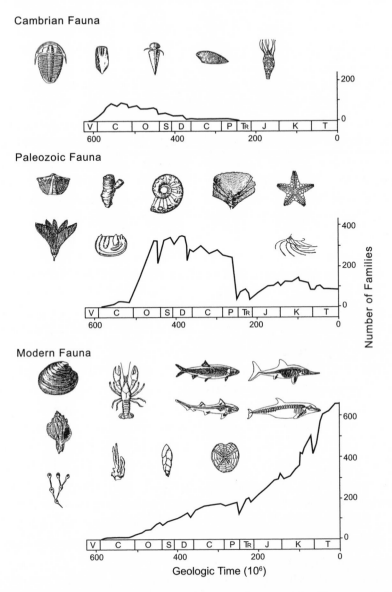

Figure 5.1 The three great marine Evolutionary Faunas of the Phanerozoic from Jack Sepkoski's classic 1984 paper. These are based on Sepkoski's database of the first and last occurrences of several thousand families of marine animals. Their diversity can be divided into three groups: the Cambrian, Paleozoic, and Modern Evolutionary Faunas. Time runs from left to right in each graph, and the letters denote each of the Phanerozoic periods, as shown in figure 1.4. Characteristic fossils of each Evolutionary Fauna are shown. From Sepkoski (1984), used with permission.

major difference between the Paleozoic and Modern groups was not ecology, but physiology, specifically in details of the circulatory systems and their level of metabolic activity. First, however, we begin by looking at the first of the twin late Permian mass extinctions, that at the end of the Guadalupian stage of the Middle Permian. Recall that we know little of this event 8–10 million years before the massive extinction at the very end of the Permian, other than that it was a major biotic crisis.

$$\cdot \quad \cdot \quad \cdot$$

Ten years ago I believed that the extinction was a lengthy affair, beginning with the retreat of the oceans at the end of the Guadalupian drying out the west Texas Permian Basin, the Zechstein Sea in northern Europe, and basins across central Asia. The extinction gradually accelerated over perhaps 10 million years to a crescendo at the end of the Permian: A long, drawn-out extinction far different from the sudden decapitation of the dinosaurs at the Cretaceous-Tertiary boundary.[2]

I was wrong.

In 1994 two groups published compelling arguments for two discrete mass extinctions that had occurred during the Late Permian, one at the close of the Guadalupian stage, and a second at the close of the Permian itself.[3] Jack Sepkoski's earlier analyses of global diversity had hinted at two extinctions, but it had been difficult to tell whether there were two discrete events, or whether a poor late Paleozoic fossil record smeared out a single event.

This backward smearing of a catastrophic event to make it look gradual is a reality of the fossil record. As a thought experiment, imagine that a mass extinction was occurring today (as, in fact, it is). Common and abundant species of clams, for example, are readily preserved as fossils. So to a paleontologist millions of years in the future it would appear as though they had survived up until our notional extinction today. But a less common species of clam may not have had any individuals join the fossil record for 1,000; 10,000; or even 50,000 years. So although this less common clam is alive today, in the fossil record it would appear as though it became extinct long ago. The common species will be preserved

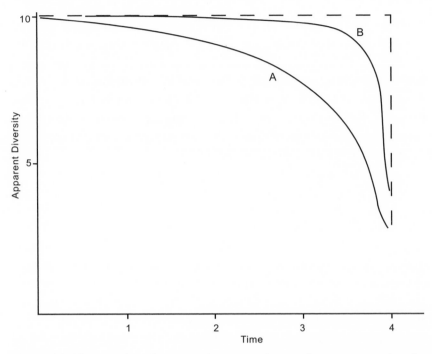

Figure 5.2 The vagaries of preservation mean that we do not expect that the very last individual of a species will be preserved and recovered from the fossil record. The more abundant a species is, and the more records we have of it, the more the last recorded discovery of the species will approach the actual last occurrence. The dotted line shows a hypothetical actual extinction curve, with B the apparently gradual pattern of a well-preserved fossil group, and A the even more gradual pattern of a poorly preserved group. This artifact means that even catastrophic extinctions will appear gradual, or smeared out in time, a pattern paleontologists know as the Signor-Lipps effect. Adapted from Signor and Lipps (1982).

right up to today (or close enough) but the last recorded occurrence of the less common forms will be smeared backward in time (figure 5.2). So even though the extinction (in our thought experiment) was catastrophic, it will necessarily appear gradual in the fossil record.[4] Today this is known as the Signor-Lipps effect after Phil Signor and Jere Lipps. The difference in the likelihood that an individual will become a fossil depends on their abundance and how easy they are to preserve. Clams, snails, and other animals with a hard, durable skeleton tend to have a good fossil record. Sea urchins and many crabs fall apart soon after death so their fossil record is not as good. And being in the ocean helps too, since to a first approximation, everything on land is being eroded and dumped into the sea.

Steve Stanley at Johns Hopkins and Yang Xingling from Nanjing University set out to see whether sampling and preservation effects were large enough to smear out a single, end-Permian mass extinction. They looked first at the fossil record of fusulinid foraminifera, a common and well-studied group of unicellular microorganisms. Of the fifty-nine known Guadalupian genera, only fourteen survived into the Late Permian and each of them was small. When they looked more closely at the data collected by Jack Sepkoski, Stanley and Yang realized that the patterns of extinction in the Guadalupian were very similar to the latest Permian. Yet if the Signor-Lipps effect had been significant, there should have been higher Guadalupian extinctions among relatively poorly preserved groups since they should have been more susceptible to preservational problems. But they found no bias. The intervals in Sepkoski's data are fairly coarse, so it was difficult to exclude the possibility that extinctions had started in the Guadalupian, and continued right up until the end of the Permian, as I originally argued. But Stanley and Yang also looked at records on the occurrences of brachiopods in China, and these clearly showed a pulse of extinction at the Guadalupian-Lopingian boundary, then a drop in extinctions until the very end of the Permian (figure 5.3).

China has a more complete and a better studied fossil record of the Late Permian than anywhere in the world. Sepkoski's data was global, but there is a trade-off between global coverage and temporal resolution. Short time spans can only be resolved within small regions or at a single locality. The wealth of data from China allowed Jin Yugan and his students to look at the Late Permian fossil record in detail, and they found a similar pattern to Stanley and Yang: two pulses of extinction, one at the end of the Guadalupian, and another at the end of the Permian.

Ammonoids are one of the few groups where both patterns of loss of taxa as well as changes in shape have been studied across the Capitanian mass extinction (figure 5.4). About two-thirds of ammonoid genera died out at the end of the Guadalupian, but dozens of new genera appeared during the Wuchiapingian and the Changhsingian. Quantitative analysis of the shape of the am-

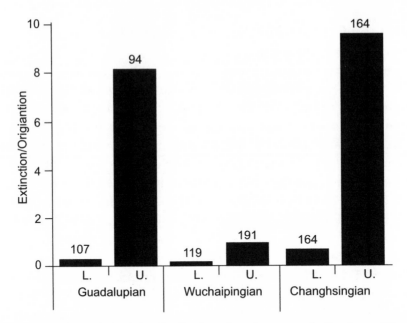

Figure 5.3 The analysis of the distribution of marine genera by Stanley and Yang clearly shows the presence of two mass extinctions during the Permian, one at the end of the Guadalupian Stage of the Middle Permian, the second at the end Changhsingian Stage of the Late Permian. Reprinted with permission from Stanley and Yang (1996). Copyright AAAS.

monoids reveals a very selective pattern of extinction, with considerable loss of forms around the margins of the morphospace. In other words, the extinction clustered the remaining shapes around the average. The expansion of new species during the Wuchaipingian occurred without the production of many new morphologies.[5]

Remarkably, a decade later we still do not know much more about this first pulse of extinction. Sequences of rock spanning the boundary are very few, almost as rare as those crossing the Permo-Triassic boundary were once thought to be. Far and away the best record of the end-Guadalupian extinction also comes from China, near the village of Penglaitan. This long sequence in southern China along the Hong Sui River is about 30 km from the Metan section discussed in chapter 3. Here we find a massive change in carbon isotopes coinciding with the mass extinction and a thick sequence of fossils. Studies of brachiopods, ammonoids, and gastropods all confirm that considerable extinction

Figure 5.4 Some of the groups that suffered heavy extinction during the end-Guadalupian mass extinction include blastoids, crinoids, tabulate corals, ammonoids, and fusulinid foraminifera. Drawings from Wanner (1931, 1937) and Girty (1908).

happened near the Guadalupian-Lopingian boundary. There are hints from our work in South Africa that there may have been an equivalent extinction among land animals as well, although this is far from certain.

The causes of this event are as obscure as the boundary is rare. The case for a global marine regression seems far more certain for the end-Guadalupian than for the end-Changhsingian. A combination of reduced shelf area for marine animals and climatic changes is at least a plausible scenario.[6] In the absence of any more definitive data about the time or nature of this episode, I will leave it to focus our attention on the end-Permian.

· · ·

Bob Linsley, my undergraduate paleontology teacher was fond of yelling out: "Stop! *Mucrospirifer*!" from the front seat of the van as it hurtled past an outcrop. The shocked student driver would shudder to a stop, Bob would climb out, bottle of Coke in hand, walk back to the outcrop, and, sure enough, pick up a *Mucrospirifer* (or whatever other brachiopod he had called out). The students would be suitably amazed at Bob's keen eyesight (particularly in view of his decidedly thick glasses) as he regaled them with some esoteric point about life during the Devonian. We would climb back into the vans, even more convinced of Bob's omniscience. By the time we were juniors we were on to Bob and were happily co-opted in his practical jokes. While I have no doubt of Bob's general omniscience (even today), he knew he could find a *Mucrospirifer* anywhere he pulled the trick.

There must be some limit to the variety of ways of putting together two shells and pumping water through them, but if so, brachiopods have yet to discover it (figure 5.5): tall cones with flaps; round flattened discs with pop-up filters; long-winged shells; fat, massive hemispheres; and the terebratulids, whose shape like a genie's lamp provides the common name of lamp shells for the phylum Brachiopoda. The variety of brachiopod sizes and shapes is remarkable given that in some sense they all do the same thing: filter feeding by pumping water while lying immobile or attached to the sea floor.

Figure 5.5 Permian brachiopods exhibit a surprisingly wide range of different shapes. These are all representatives from the collections of the National Museum of Natural History.

Brachiopods are as characteristic of the Paleozoic as dinosaurs are of the Mesozoic. They have a pair of hinged shells, similar to a clam, but are unrelated to clams, or indeed to any mollusc. Clam shells are generally mirror images of each other, like left and right hands. But in brachiopods the plane of mirror symmetry is at right angles to the plane between the two valves of the shell. So the left half of one valve is the mirror image of the right half. Inside this almost cosmetic difference lies a very different architecture. Brachiopods are exquisitely well adapted to filtering food particles by pumping seawater through a coiled curtain of filaments attached to a U-shaped loop. At the base of the loop lies the mouth, and the filaments feed particles down toward the mouth. Most articulate brachiopods, the major class, are extinct and comparatively uncommon today.

About 90% of brachiopod families and genera disappeared between the mid-Permian and the Early Triassic. Guirong Shi and Neil Archbold of Deakin University in Australia have spent much of the past decade scouring old paleontological papers to compile records of each Permian brachiopod occurrence, with particular emphasis on the western Pacific. There are gaps in the data where paleontologists have not been active, but these compilations do illustrate where brachiopods were distributed, and when plotted on maps of Permian geography, they reveal connections between continents, the flow of oceanic currents, and even insights into climatic patterns. Shi's group recorded 141 genera in fifty families during the Changhsingian Stage in an area stretching from the Mediterranean through the Middle East and into Asia and Australia. Although the numbers are less than earlier in the Permian, it was still a diverse assemblage. The region occupied by these tropical brachiopods appears to have expanded toward the end of the Permian. Shi could find no records of brachiopods in high northern areas, suggesting they disappeared before the end-Permian extinction. Brachiopods persisted in the high southern latitudes of New Zealand and Australia. The Deakin University group explained the changing distributions of brachiopods as a result of Late Permian global warming leading up to the extinction.[7]

An old paleontology textbook, describing the end-Permian extinction of brachiopods would mark the passing of the orthid brachiopods, major losses among strophomenids, and lesser losses among the wing-shaped spiriferids, but only minor losses among the rynchonellids and tererbratulids. These wonderfully named beasts represent the classically defined major groups of articulate brachiopods, committed to memory by generations of geology students and prized by fossil collectors.

No need to memorize them, however, for most do not exist, at least not in any real sense. Each of these groups was defined on the basis of overall shape, the structure of the internal supports for the complexly folded, filamentous lophophore, the nature of the hinge between the valves, and a few other features. Over the past several decades systematists, the scientists charged with unraveling familial relationships between organisms, have adopted a more explicit and rigorous approach. This cladistic approach involves identifying specific features shared between organisms as the basis for establishing patterns of evolutionary relationships. To the shock of all (and the disgust of some, it must be admitted) many groups reveal a very different pattern of relationships than previously thought, and for a fascinating reason. Just as wings are found among birds, insects, bats, and the extinct pterosaurs, many other features of animals evolved multiple times. The explicit approach of cladistics provides a more rigorous approach to identifying these patterns of convergent evolution. The wings of birds, insects, bats, and pterosaurs are nothing alike and there has never been any question that wings and flying evolved independently in each group. Careful study of museum specimens and sophisticated statistical tools are required to recognize such convergences in other features. Convergence turns out to be so ubiquitous that many long-accepted combinations turn out to represent unrelated genera and species. It is as if one claimed that all the girls dressed like Britney Spears or men in gray suits were part of the same family.

This may seem like a tedious disagreement among pedants (and, in truth, some of it is), but the pervasiveness of convergence is a testament to the incredible power of evolution to craft solutions. Many solutions are so effective that natural selection has rediscovered them time and time again, generating the same shape or similar structures to common evolutionary problems. Systematists are only able to unravel the patterns of relationship by carefully scrutinizing each aspect of a species and not privileging any particular part.

Sandy Carlson at the University of California at Davis has carefully unraveled these complexities among the brachiopods, showing that many of the traditional groups of brachiopods represent forms that evolved multiple times. Terebratulids, the group that looks like a genie's lamp, do apparently represent a single group, or clade. The long spiriferids, in contrast, evolved several times. All of this complicates interpreting the pattern of extinction across the boundary. I will desist from describing patterns of extinction among the larger subgroups of brachiopods. This same problem extends to many other groups as well. It is slowly getting sorted out as new systematists tackle these groups, and as more sophisticated methods of data handling are introduced.[8]

Brachiopods are closely related to another phylum, the bryozoa. Bryozoans live in colonies, generally in a well-protected skeleton, with the tiny individuals popping their filtering fan out of a hole. Individual colonies are well integrated, building large frameworks optimized for water flow through the colony. The lacy fenestrate bryozoans were common inhabitants of Permian reefs, along with sponges, corals, brachiopods, and crinoids. The stenolaemate bryozoans, the most significant clade of Permian bryozoans, began a steady decline in diversity early in the Middle Permian, well before either of the two major extinctions. Late Permian bryozoans are relatively rare, with a few reports from the tropics, including Iran and Pakistan, but none from China. More study is needed but the major pulse of extinction apparently occurred during the Guadalupian, with the few surviving species disappearing at the end of the Permian. In contrast, the gymnolaemate bryozoans, ancestors of modern groups, sailed through Late Permian events essentially unscathed.

Crinoids, or sea lilies, were the third major component of Permian communities and the Paleozoic Evolutionary Fauna. Attached to the sea floor via a long stalk of a stack of disc-shaped columnals, crinoids were topped by a cup with feathery arms surrounding the mouth. The arms catch tiny prey. Although living crinoids are often free-living, the stalked Paleozoic crinoids formed great stands, with different species at different heights, like the shrubs, bushes, and canopy trees of a modern forest. Today the echinoderms we know are starfish and sea urchins, and some may recall the sea cucumbers along the beach that dump their guts out their anus when prodded. (An effective strategy for distracting predators while the sea cucumber crawls away, such self-evisceration also attracts small boys to pester a multitude of benign animals.) Echinoderms, the phylum encompassing crinoids, their cousins the blastoids, star fish and sea urchins, has only five living classes, but had as many as twenty-one during the Early Paleozoic. Despite the ubiquity of modern sea urchins, echinoderms are a shadow of their earlier diversity.

The tabulate and rugose corals were another important part of the Paleozoic fauna, and also disappeared at the end of the Permian. The modern corals, the scleractinians, first appeared either in the Middle Triassic, or possibly the latest Permian, but represent an independent origin of corals from some unskeletonized sea anemone. My friend Wang Xiang-dong, a student of Jin Yugan's, recently studied about 2,100 coral genera from the Permian of China. Although Late Permian corals remained diverse in China, and perhaps elsewhere in the tropics, they disappeared from the Arctic by the Changhsingian, much as brachiopods did. Elsewhere, however, latest Permian corals are found in Spitsbergen and Greenland. Fossils can be scarce in the Changhsingian rocks outside of China, so this apparent geographic difference in extinction may say more about where paleontologists have found fossils than about the pattern of extinction, but it remains an intriguing issue for future research. Although both tabulate and rugose corals were almost obliterated by the end-Guadalupian mass extinction, ten families and 107 species persisted into the Late Permian, but all became extinct at the boundary.[9]

Brachiopods, bryozoans, stalked crinoids, and corals shared a way of making a living: all lived attached to the sea bottom and filtered food out of the water. Some groups were remarkably selective about what they filtered, straining out just the right particles and discarding the rest. The abundance (number of individuals) and diversity (number of species) of filter feeders from the Ordovician through the Permian testifies to the success of this strategy. Carboniferous limestones composed almost entirely of broken crinoid stalks litter vast sections of the American Midwest. Almost one thousand different species of brachiopod have been described from the reefs of the Permian Basin in West Texas, the Permian equivalent of today's Great Barrier Reef in Australia. These groups thrived in similar circumstances. Mud is not particularly nutritious and requires energy to remove. Attachment required a firm bottom without the bulldozing effect of burrowing clams, crabs, and the like disturbing the sediment. And the wealth of these filter feeders required an enormous and ongoing supply of nutrients. High sea levels during much of the late Paleozoic provided broad, tropical shelves for these groups to flourish. Something about this way of making a living exposed them to greater extinction at the close of the Permian. The collapse of these groups could reflect the disappearance of the food source they shared, the clear water they required, or the stable substrate beneath them. But disappear they did.

· · ·

The Modern Evolutionary Fauna are the survivors. Some individual groups suffered high extinction, but overall extinction was low. The few Paleozoic predators belonged to these groups, and from them would come the very different ecosystems of the post-Paleozoic. Only one or two species of the eight Permian genera of sea urchin survived the mass extinction, but they have been tremendously successful during the past 200 million years. Ammonoids and other molluscs suffered less extinction but display some interesting diversity patterns, so it is worth considering each group in greater detail.

Ammonoids were a prolific if unstable group of cephalopods. Outwardly similar to their cousins the nautiloids and *Nautilus*, the coiled ammonoids had far more complex patterns of sutures, the attachments between the outer shell and the septa that divide the shell into chambers. As in *Nautilus*, the chambers were presumably used to regulate their buoyancy, and the great complexity of the sutures may indicate that these active predators could pursue their prey to great depths. Ammonoids were the ultimate boom-and-bust group throughout their history. I have already mentioned their extinctions during the Capitanian crisis, but over and over they proliferated wildly, only to crash ignominiously during the next biotic crisis.

The dynamism of ammonoid history probably contributed to their eventual demise during the end-Cretaceous mass extinction, and it almost brought about their disappearance during the Late Permian. The ammonoid brain trust, comprised of the world's leading experts on Permian ammonoids, recently collated records from around the world and revised many of the taxonomic assignments. They concluded that some 97% of all ammonoid genera alive in the Changhsingian vanished by the end of the period. The analysis of changes in ammonoid shape revealed that unlike the Capitanian crisis, the pattern of extinction was not concentrated on particular shapes, but impacted the whole suite of morphologies.[10]

Turning from the predatory ammonoids to the filter-feeding bivalves, we come to one of the enigmas of the extinction. If filter feeders were generally selected against, how did bivalves survive? True, some clams did other things, but most Permian clams were either suspension feeders, like brachiopods, filtering small prey from water or were deposit feeders that screened detritus out of the mud. Many modern clams, including goey-ducks, have two fleshy siphons—one to take in water, and another for the return flow. Siphons free clams from living on the surface, allowing them to burrow deep into the sediment for both protection and in search of food. Fused siphons first appeared in the Permian but clams had yet to make great use of them. Although many scallops and a few other groups of clams became extinct, overall only 26% of all families disappeared in the Late Permian. The relative success of clams suggests that something other than just filter feeding was responsible for the differential patterns of extinction and survival.

The gastropods, or snails, are my favorite group. With several postdoctoral students and other colleagues I have spent much of the past two decades collecting Permian snails, sorting through their family history, and trying to reconstruct what happened to them during the extinction (figure 5.6). Some 20%–25% of gastropod genera disappeared from the Middle Permian through the Wuchaipingian, but 41% of genera disappeared at the close of the Changhsingian. Although there were relatively few extinctions across the group as a whole, the mass extinction was a watershed in their history. Today about 90% of all marine snails are carnivorous, and most are predatory. These range from the tropical cone shells with sophisticated neurotoxins to naticids boring into the shells of other gastropods. To be sure, abalones and the high-spired turritellids remain, feeding on detritus and suspended matter, but this feeding strategy is far less common today than during

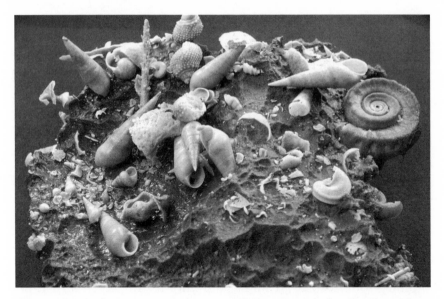

Figure 5.6 Silicified gastropods from west Texas, showing the diversity of Late Permian snails. This photograph was taken by my predecessor studying Permian gastropods at the National Museum of Natural History, J. Brooks Knight, in the 1950s.

the Paleozoic. This change in events comes from the tremendous increase in the diversity of snails during the past 250 million years.

Permian gastropods had many different ways of making a living, broad geographic distribution, and moderate extinction. This makes them an ideal group for studying the causes of the extinction. This may seem paradoxical, but the destruction of brachiopods or ammonoids was so complete that it makes it difficult to say anything useful about the actual cause of the extinction. The essence of solving the problem of differential extinction is being able to compare similar winners and losers. A clade where all species survived is not particularly edifying; nor is a group that almost completely disappeared. But with bivalves and gastropods there are enough of both winners and losers to make useful comparisons. While there is still much more work to be done among gastropods, we have been able to separate out genera with broad geographic distribution from those found in single biogeographic regions. The end-Guadalupian mass extinction looks much like the non-mass-extinction intervals on either side: widespread genera were far less likely to become extinct than those with narrow distribution. Things seem to reverse at the end of the Changhsingian, when broadly and narrowly distributed genera are about equally likely to disappear. We know little about whether some areas of the globe suffered greater extinction than others, but some of our data on gastropods suggest that the extinction was more severe in Europe and North America than in China and adjacent parts of Asia.[11]

. . .

The destruction of modern coral reefs is a harbinger of the general decline of ocean health. Paleontologists have long had similar views of reef ecosystems as fragile and easily destroyed communities. Ecologists study reefs as a primary example of a tightly integrated community, where species are highly connected and the disappearance of one species can cause the cascading loss of many more. This view of reefs as tightly connected and even fragile superorganisms gains support from claims that reefs are one of the first ecosystems to disappear during mass extinctions. Today we think of reefs built by corals and teeming with brightly colored fish. But reefs have a history dating back over 2 billion years to massive carbonate structures built by microbes along an ancient coastline in northern Canada. Microbes, sponges, corals, algae that secrete calcareous plates, massive bivalves, and tiny worms have all built subsequent reefs. Linking these reefs is an environment where organisms can produce large volumes of carbonate. Some reefs grow in very deep water, but here I will restrict my focus to reefs in shallower water, probably in bright sunlight.

During the Middle and Late Permian, extensive reefs developed in west Texas, Greenland, northern Europe, the Mediterranean and Oman, and east through the Caucasus mountains, Pakistan, Iran, and into China and southeast Asia. Unlike modern reefs, however, microbes, algae, big sponges, bryozoans, and even brachiopods played significant roles in their construction. There are at least ten different types of Permian reefs, with those made by algae the most common.

Reefs built by calcareous algae disappeared during the Guadalupian extinction event as the West Texas and some other basins dried out. The disappearance of reefs explains the loss of many sponges. But reefs built by microbes, other sponges, and corals persisted into the latest Changhsingian in Greece, Tunisia, and through many places in South China. There is little doubt that reefs continued hale and hearty until the very end of the Permian in several parts of the world, with no diminution through the Changhsingian. Erik Flügel and Wolfgang Kiessling at the Institute for Paleontology in Erlangen, Germany built a massive database documenting 600 million years of reef history. Their data reveals the magnitude of reef destruction during the Late Permian. Across the Permo-Triassic boundary, the number of reefs alive per million years and the rate of carbonate production drop to zero, the mean thickness of a reef plummets 96%, and the diversity of reefs declines by 80%, to levels equaled only during the end-Ordovician and end-Cretaceous mass extinctions.

Reefs commonly develop on broad carbonate shelves where light and nutrients are abundant. Reefs do not grow near the mouth of the Amazon because the mud and muck emptying out of the river would clog the filters of suspension feeders and prevent light from reaching the photosynthetic algae growing in the tissues of corals. In her recent history of reefs, my friend Rachel Wood from Cambridge University pointed out that there has been remarkably little rigorous testing of the claim that reefs are the first to suffer during mass extinction, and the one reliable study found no evidence for higher extinction rates among reef organisms. Since carbonates do not develop without a platform upon which to build, Rachel reasoned that the disappearance of reefs during mass extinctions could reflect the loss of the carbonate platforms as much as the loss of reef animals. With our focus on animals, we tend to assume that they drive ecosystems, but the environment can often play just as significant a role. To test this idea we need to separate the study of reefs from the vicissitudes of the organisms that build them.[12]

The Flügel and Kiessling reef database largely confirms Rachel's intuition. Across the Permo-Triassic boundary as reefs disappeared, their data show that the production of carbonate sediment declined dramatically. Flügel and Kiessling uncovered no evidence that reefs retreated toward the tropics during the end-Permian crisis. The destruction of reefs is clear, and unlike the earlier Guadalupian extinction when sea level dropped, it does not seem that we can invoke loss of habitat area as a cause. So we are left with an unresolved conundrum. The drop in the rate of carbonate production surely had a catastrophic effect on reefs, but was that the cause or effect? Certainly reef-dwelling animals disappeared, and production of carbonate across carbonate platforms also collapsed, but the most likely explanation may be that each reflects a broader, more encompassing cause.

"Absence of evidence is not evidence of absence" sounds like a bad line from the trial of O. J. Simpson, but applies to the continually perplexing issue of differences in extinction across latitude. Paleontologists' predisposition toward assuming higher extinction rates in the tropics relative to high latitudes may say much about paleontologists but little about extinction. In this chapter I have noted claims of greater extinction among high-latitude brachiopods and corals, and Flugel's rejection of such claims for reefs. Only through careful sampling of different regions can we discern whether geography played a role in the extinction, but paleontologists have yet to work this out for the Permo-Triassic event. The few hints we have are enough to justify some intrepid graduate students pursuing the issue.

. . .

Paleontologists prefer to think about bones, shells, and skeletons, the biotic debris that make good skeletons. Physiology, the metabolic interactions and cycles of living organisms are a less common concern, at least for invertebrate paleontologists. Thus one of the most innovative views of the extinction (innovative in the positive and creative sense, as opposed to innovative in the polite sense of utterly divorced from reality) is the suggestion that patterns of extinction and survival were driven not by latitude, way of making a living, or number of species but by physiology. Before Richard Bambach, Andy Knoll, and their colleagues ginned up this new approach to sorting Jack Sepkoski's diversity data, paleontologists had rarely employed presumed physiological attributes of long-dead groups to explain diversity.

Bambach and Knoll divided most marine animals into two groups: the first with a heavy, calcified skeleton, absorbs oxygen through tissues but without gills and with weak or passive circulation systems. This group generally has lower metabolic rates and includes the tabulate and rugose corals, stenolaemate bryozoans, brachiopods and the stalked echinoderms, blastoids and crinoids. The second group actively acquires oxygen through gills, has an active circulatory system and high metabolic rates. Many molluscs, the arthropods, conodonts, and vertebrates fall within this category. They view the low metabolism group as essentially open or unbuffered against environmental stress, particularly those induced by changes in ocean chemistry. The higher metabolism, active circulation and well-developed respiratory systems of the second group constituted a closed and buffered system capable of adjusting more readily to chemical insult. The insult, they proposed, was very high concentrations of carbon dioxide in the water that overwhelmed the ability of animals with low metabolic rates to deliver sufficient oxygen to their tissues.[13] During the end-Guadalupian extinction the low-metabolic genera suffered 65% extinction, versus 49% for the second group. Similarly for the Changhsingian, 81% of the first group disappeared but only 38% of genera in the high metabolic rate clades.

This parsing by metabolic capacity is the first novel view of extinction selectivity in decades. Some issues remain unresolved, particularly whether the twofold division of physiological attributes is the most appropriate way of treating the data or whether the diversity of metabolisms, gill structures, and skeletons demands a more subtle approach. Richard Bambach, Andy Knoll, and Jack Sepkoski followed up this initial foray into paleophysiology with a second study chronicling the history of groups with these different physiologies from the Ordovician radiation up to today. During the 200 million years leading up to the end-Guadalupian mass extinction, the unbuffered groups were twice as diverse as the buffered ones. The extinctions destroyed this pattern but the success of the buffered forms did not last. While 60% of Early Triassic genera fall within the buffered category, unbuffered groups diversified more rapidly, and by the Early Jurassic, the diversity of the two groups had equilibrated and remained so until the end-Cretaceous mass extinction.

The three possibilities—ecology, association with reefs, and physiology—are not completely independent, and further study may indicate synergistic interactions between them in promoting survival or enhancing extinction. While the paleophysiology approach of Knoll and Bambach requires more clarification, it does provide the best explanation of the extinction pattern. The role of filter feeding may have been an important factor in the extinction, mediated through a loss of zooplankton and phytoplankton, but it does not seem to explain the diversity patterns quite as well as physiology. The history of Permian reefs is complicated, both by the number of different kinds of reefs, and by the uncertainty over whether reefs were specifically targeted or disappeared because of a loss of habitat. Rising sea levels make the latter unlikely, so we must seek other explanations for the catastrophic disappearance of reefs.

· · ·

The final issue to consider is the rapidity of the marine extinctions. In chapter 3 I introduced our data from the Meishan section showing that the extinction appeared sudden. But how sudden is sudden? I want to return to that data, and integrate it with the radiometric age dates from chapter 4 to discuss the rapidity of the extinction. One lesson of the Signor-Lipps effect is that paleontologists cannot necessarily believe the pattern of fossil disappearances pulled out of the rock. This has been a difficult lesson for some paleontologists. Remember that the Signor-Lipps effect suggested that even catastrophic extinctions would appear gradual in time because of the vagaries of fossil preservation. Abundant and common species, those recorded from every layer have a far better record than those found only occasionally. The last recorded appearance of the common species must be much closer to the level of the actual extinction than the last of a rare fossil (figure 5.2). In the aftermath of the Alvarez paper on an extraterrestrial cause for the end-Cretaceous mass extinction, a series of meetings have brought together scientists working on mass extinctions, whether physicists with experience in impact dynamics or paleontologists who knew the rocks and fossils. At the third of these affairs a talk on the Signor-Lipps effect and its implications was interrupted by a red-faced paleontologist bellowing: "I sweated blood collecting each of those fossils and I'll be damned if you are going to tell me I can't use them!"

If the Greek gods were with us still, some minor sprite in the pantheon would be charged with mucking up the fossil record to mislead paleontologists. But my colleagues' protestations are misplaced, for the pattern of fossil disappearances can mislead us. The solution lies in the fossil records themselves. By using the number of occurrences of a fossil leading up to the extinction we can begin to place some statistical estimates on where the true last appearance occurred.

The Meishan section is the only place in the world where the fossils across the Permo-Triassic boundary have been studied intensively enough to apply this approach. As I discussed in chapter 3, we have a record of each occurrence of 333 different species in 162 genera through the 80 meters of Changhsingian and Early Triassic strata. We collapsed 265 species into 93 genera and then added the radiometric dates from the various ashes as discussed in chapter 4. This allowed us to essentially rescale the data by time, rather than thickness of the rock (figure 5.7), and we were able to show that the 93 genera that disappeared near the boundary were consistent with a sudden extinction at beds 25 and 26. We looked at the data in several different ways as well, but the conclusion stands, that there was a single extinction event, coincident with the shift in carbon isotopes and the volcanic ashes.[14]

The Meishan fossils clearly demonstrate an abrupt extinction, although how abrupt is still a matter of debate. Meishan is only one very tiny spot on the globe, and while it looms very large in discussions of the end-Permian mass extinction, and in this book, there is no certainty that what we learn there holds true anywhere else. The Permo-Triassic boundary sections in the southern Alps have been studied in as much detail as some of the Chinese sections, generating a robust amount of data, and also provide some insight into what was happening on the supercontinent of Pan-

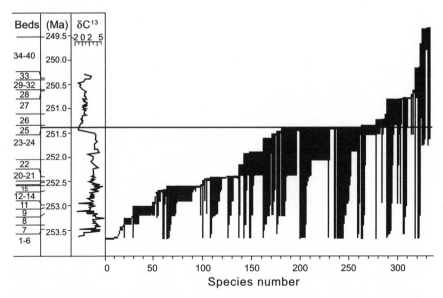

Figure 5.7 Distribution of genera at Meishan, as in figure 3.5, but rescaled to time based on the radiometric dates. From Jin et al. (2000).

• • •

gaea. Although far fewer fossils are involved, the distribution of thirty species of foraminifera follows the pattern expected of a sudden extinction, and is unlike that for either a stepwise or gradual extinction.[15]

• • •

Did crinoids and brachiopods disappear at the end of the Permian because they simply lost the battle for survival to the better adapted, or more nefarious, claims and snails? David Raup calls this the Fair Game scenario. Raup, one of the intellectual giants of paleontology, distinguishes three classes of mass extinctions.[16] The Fair Game scenario is normal, Darwinian extinction in which the best-adapted species survive and more poorly adapted species disappear. A Field of Bullets extinction is akin to the Battle of Gettysburg during the Civil War, with a hail of bullets randomly extinguishing individuals rather than species. Probability takes over, and species with fewer individuals are likely to go extinct while abundant and widespread species are likely to survive. Wanton Destruction is not the latest Hollywood toys-for-boy's movie, but an extinction in which the rules change and those features that contributed to success before an extinction no longer work. While no mass extinction is likely to exactly match any of these scenarios, they serve to focus our attention on the long-term impact of these extreme events.

The Wanton Destruction scenario is the most interesting to paleontologists and evolutionary biologists because it sorts species via fundamentally different rules than those that operate during normal Darwinian evolution. Long-established adaptive trends may be of little survival value, and apparently well-adapted groups may simply disappear. Severe economic recessions often have a similar effect on jobs and the survival of companies. A company with good, prudent managers, a valuable product, and productive workers may disappear during a sudden economic reversal simply because credit is suddenly withdrawn, suppliers fail, or creditors pay too many bills late. In contrast, if the Fair Game scenario is operating, a mass extinction may speed up preexisting patterns of gain and loss of species. Trends may be achieved more rapidly. This cuts to the heart of our understanding of the history of life, for if Wanton Destruction is common, established evolutionary trends might have little influence on longer-term trends in the history of life.

Fair Game, Wanton Destruction, and Field of Bullets. Choosing between Dave Raup's alternatives requires knowing both the abundance of individuals within species as well as the changing patterns of the number of species. Unfortunately we have little information on the abundance of fossils. Gathering this sort of information requires a well-thought-out sampling scheme to ensure that comparable material is acquired from different areas. Few such collections exist for any mass extinction, and I can think of none for the end-Permian event. We do know that the extinction was selective, decimating some groups while sparing others. As I have discussed in this chapter, the differences in metabolism between the open, unbuffered Paleozoic Evolutionary Fauna, and the closed, buffered metabolisms of the Modern Evolutionary Fauna suggest that physiology may have been more critical than living on a reef, or ecology.

Many of the proposed causes of the extinction apply only to the oceans. Anoxia and changes in sea level, for example, might be effective causes of marine extinction, but if extensive extinctions occurred on land as well, we must look beyond such ideas to explain the extinction. So in the next chapter we turn to events on land, focusing on perhaps the best-studied terrestrial area, the stark but beautiful Karoo of South Africa.

CHAPTER 6
South African Eden

Farm Meltonwold, near Victoria West, Karoo Desert,
South Africa March 2002

As we came around the low rise, the skeleton was easy to spot from a distance, an eight-foot-long mound of bones standing above a gray mudstone. This was a typical *Bradysaurus* skeleton: complete, all four legs planted, looking like the animal died during a nap (figure 6.1). Contrast this to the fate of a skeleton in Africa today: Lions, hyenas and other scavengers tear apart even an adult elephant in a few days. Rare is the modern vertebrate skeleton that makes it into the fossil record unmolested. Yet *Bradysaurus* seems to have suffered a different fate. Some paleontologists have suggested they were trapped in mud during floods and could not extricate themselves, much as Ice Age mammals were trapped by oil sands in the La Brea Tar pits a few thousand years ago. This fossil pareiasaur surrounded by a sea of gray mudstone confirms that part of the hypothesis, but I still wonder that there are so few signs of scavenging among the Karoo fossils.

In this chapter we turn from the sea, our concern in the past few chapters, back to the land. By the Late Permian the first stubby plants of the Silurian and Devonian had evolved into a vast array of different groups including conifers and seed ferns.

Figure 6.1 The *Bradysaurus* skeleton at Farm Meltonwold.

Flowering plants had not yet evolved but would hardly have been missed among the towering conifers and other trees. Arthropods emerged from the sea, and by the Late Permian every major order of insects had appeared, along with a number of now extinct groups. The first vertebrate animals crawled onto the land during the Late Devonian, some 375 million years ago. Amphibians begat reptiles, which split into several groups by the Permian, with one lineage leading to mammal-like reptiles and eventually the mammals, and another lineage leading to reptiles, including dinosaurs, birds, lizards, and snakes. Mammals, dinosaurs, and turtles would not appear until the Triassic, but the land was well stocked by the Late Permian with a wonderful zoo of dicynodonts, gorgonopsids, and therapsids in addition to pareiasaurs like *Bradysaurus*.

Critical to understanding the events of the Permian mass extinction is establishing whether extinction occurred on land. For decades paleontologists assumed that the mass extinction was entirely a marine affair, with little impact on terrestrial ecosystems. This assumption greatly influenced various extinction scenarios, many of which would have little discernible impact on life on land. How did plants, vertebrates, and insects fare across the boundary? As in the oceans there are significant extinctions among the vertebrates, a major impact on plants, and insects suffered what may be the only mass extinction in their history. The sediments preserved on land have an equally important tale to tell of sudden changes in climate at the Permo-Triassic boundary. Finally, we need to establish whether events on land happened at the same time as those in the sea; if they did, all well and good, but if not, we would need to concern ourselves with a more complex interaction between land and sea.

• • •

We met Bruce Rubidge in his bakkie at the Shell Station on the north side of Beaufort West, a small town on the edge of the Great Karoo Desert of South Africa. (South Africans, like Australians, speak a dialect clearly, if distantly, related to English; *bakkie* is South African for a small pickup truck.) Sam Bowring and I had flown to Cape Town two days before, met Sam's colleague Maarten de Wit from the University of Cape Town, and headed into the field. Bruce and I had met at a party in the Museum of Natural History in Beijing several years earlier, and he had encouraged me to visit the Permo-Triassic boundary sections of South Africa, but it was Sam's longtime collaboration with Martin that finally gave me a chance to take Bruce up on his invitation. I knew Bruce as the director of the Bernard Price Institute for Palaeontological Research at the University of Witwatersrand in Johannesburg, and as one of the foremost vertebrate paleontologists studying Permian and Triassic vertebrates, but I am chagrined to admit how little I knew of his family's long history in the Karoo as pioneers of that arid but beautiful desert, or their pivotal role in discovering the wealth of vertebrate fossils that has made it a mecca for paleontologists for some 150 years.

As we headed north into the Karoo, I gradually lapsed into silence, partly from the effects of the fifteen-hour flight from Atlanta, but mostly in wonder at my first view of the Karoo. There is nothing like a good desert to make me happy. I spent most of my childhood summers among the mesas and scrublands of the American Southwest. The red rocks, mesas (here called *kopjes*), and barren scrubland looked just like home to me (figure 6.2).

Figure 6.2 The Great Karoo of South Africa, from atop Lootsberg Pass, looking out across the Karoo. Permian rocks are at the base of the hill while Jurassic basalts cap the ridges in the distance on the left.

The Great Karoo is a broad basin covered with scraggly bush and occasional thorn trees that stretches across the middle of South Africa from Cape Town in the southwest toward Johannesburg in the northeast. A thin strand of highway and an increasingly decrepit rail service cuts through the Karoo linking the two cities. Today most visitors, and even many South Africans, explore the Karoo in air-conditioned comfort as they speed along the highway at 120 kilometers per hour. Pity, because this is a stunningly beautiful region with history encompassing the earliest hominids, traces of the San people (formerly and pejoratively known as the Bushmen), and towns built by the earliest Dutch and English settlers. Stone towers protecting bridges and critical junctions along the rail line recall the Boer War at the turn of the twentieth century.

To the south, the Karoo is rimmed by the long sweep of the 2,000-meter-high Cape Fold Belt paralleling the coast (figure 6.3). The collision with Antarctica during the Permian that created the mountains also produced a mélange of low ridges and broad grasslands in the Karoo. These undulations diminish toward the north, producing a broad, flat basin. The Great Escarpment is one of the most prominent geographic features in southern Africa, mimicking the coastline about 150 to 300 kilometers inward and dividing the Karoo basin from the highveld, a high interior plateau created by doloritic igneous rocks that formed during the initial opening of the Atlantic Ocean in the Jurassic. These tough doloritic sills are very resistant to erosion, so the edge of these rocks produces a craggy barrier along the north edge of the Karoo.

We stopped for supplies in the small town of Graaff-Reinet and then drove about twenty minutes northwest before turning left at a long line of huge aloe plants. This was Wellwood, the family farm where Bruce had grown up and that his brother Robert farms today (figure 6.4). In my ignorance of vertebrate paleontology, I did not realize then that Wellwood has long been venerated as the cradle of Karoo vertebrate studies. The farmhouse sits today as it has since the 1840s, with its magnificent whitewashed Cape-Dutch architecture, a long veranda, small green lawn and garden of flowers. Although trees are rare in the Karoo, here they tower over the house and surrounding buildings. Soon after we arrived Bruce and Robert brought out the massive commonplace book where

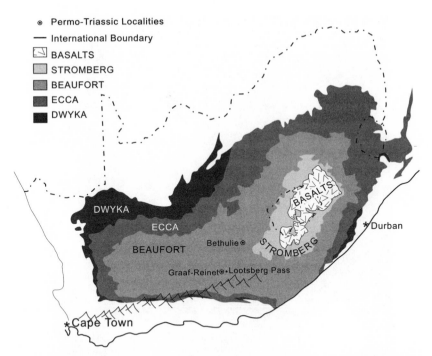

Figure 6.3 A simplified view of the geology of South Africa showing the Karoo with the Cape Fold Belt along the southern margin. The Dwyka was deposited by the Permo-Carboniferous glaciation; the Ecca is lower Permian rocks, while the interesting vertebrates are found in the middle Permian–Triassic Beaufort Group. The Lootsberg Pass and Bethulie Permo-Triassic boundary localities are indicated.

the events on the farm have been recorded each day for over a century and a half. As we settled in with drinks on the veranda we looked up the exact day on which the trees surrounding the house had been planted, when improvements had been made on the farm, and what the weather had been a century earlier.

Captain Robert Henry Rubidge and his family moved into the Graaff-Reinet district of the Karoo of South Africa in 1838. He was one of the earliest English settlers in the area and purchased the farm Wellwood for his son Charles in 1840. Over the years, five generations of Rubidges have developed it into one of the premier merino-sheep stud farms in the region. Pictures of prize rams from past decades line the rooms of the farmhouse. The Karoo is a challenging area for any farmer, cold in winter, very hot in summer, and cursed by little water. When the rains come, the grass

Figure 6.4 Robert (right) and Bruce Rubidge in the door of the collections at Wellwood Farm, home of the Rubidge family in the Karoo.

stretches for miles; in other years the burnt red earth and rocks show through. The 12,000 hectares (6,000 acres) of Wellwood stretches across a dry plain around the old farmhouse. Behind the farm headquarters a steep escarpment rises into the high country, allowing Robert and his staff to move the sheep among pastures over the course of the year.

Through dinner Bruce and Robert told us the story of vertebrates in the Karoo, and the role their grandfather played in assembling one of the world's greatest collections of vertebrate fossils. Andrew Geddes Bain, South Africa's first geologist and greatest road builder, found a few teeth and bones near Fort Beaufort in 1838. Proper British colonialist that he was, he sent them to the most noted vertebrate paleontologist in the world, Sir Richard Owen at the British Museum in London. Later that year Bain found a magnificent fossil reptile he called the Blinkwater Monster, the first example of a bizarre genus with heavy, knobby triangular skulls that would eventually be named *Pareiasaurus*. Almost the ugliest vertebrates I have ever seen, but if I studied vertebrates, pareiasaurs would be the beasts of choice.

Bain collected additional material for Owen, but the true father of vertebrate studies in the Karoo was Robert Broom, who arrived in South Africa in 1897 as part of the Scottish diaspora, medical degree in hand from Glasgow University. British doctors had replaced parsons as the purveyors of natural history to Victoria's empire, and Broom was at least as interested in natural history as in medicine. Learning of the Karoo fossils from other collectors in Cape Town, he moved to Pearston on the edge of the Karoo, set up his practice, and began collecting fossils, lizards, plants, almost everything, in fact. News travels fast in such a sparsely settled country, and particularly news of a new doctor with an interest in fossils. C.J.M. and James Kitching, two brothers in the road-building business, wrote Broom to tell him of fossils near the small settlement of Nieu Bethesda, just through the hills past Wellwood. Broom responded quickly, and his enthusiasm for the fossils spurred C.J.M. ("Croonie") Kitching's interest and the two joined forces to scour the Karoo for fossils. Most of Kitching's vast collection now resides at the American Museum of Natural History in New York. Broom eventually became a member of the Transvaal Museum in Pretoria and an internationally respected paleontologist.

According to family lore the Rubidge fossil hunting began one day during the 1930s when ten-year-old Peggy Rubidge asked her father, Dr. Sidney Rubidge (Bruce's grandfather), what a fossil was. As a younger man Rubidge had noticed the occasional fossil while riding across the veld and knew where to take his daughter collecting. Soon thereafter Peggy and a friend found bones encased in a rock and Sidney excavated them at the kitchen table with a hammer and chisel (much to his later horror!). When Broom saw them he pronounced the fossil a new species of carnivorous reptile, eventually naming it *Dinogorgon rubidgei*. Sidney Rubidge had found his dream. By 1934 he had collected over thirty new genera and seventy new species.[1]

The Rubidges are wonderful showmen in a rather low-key way. They kept us waiting until after breakfast the following morning before we walked the hundred meters down the lane to a whitewashed building and Robert ushered us into the fossil collection. In place of the long rows of white cabinets we have housing collections at the Smithsonian, everything was on display in custom-built glass cases. Sidney Rubidge's monument is one of the world's greatest collections of Karoo fossils—and the only one still in the Karoo. Today there are over nine hundred specimens there. With 118 holotypes the collections remain important to paleontologists. Holotypes are the specimens that carry the name of a new species and remain vital for comparative purposes. As we entered our names in the guest book, we thumbed back to see the names of previous visitors, from neighbors in the Karoo to some of the most noted vertebrate paleontologists, and a field excursion from the 1929 International Geological Congress in Pretoria.

I was immediately drawn to *Pareiasaurus*, one of my favorite pareiasaurs. A squat, hulking plant-eater with a sprawling lizardlike posture, the knobby, triangular head and a series of bony plates or scutes provided protection from passing predators. A young paleontologist, Michael Lee, recently suggested that these scutes gradually coalesced into a more rigid structure with the herbivorous pareiasaurs eventually giving rise to turtles in the Triassic. This union of two of the most impressive vertebrate clades should be true on purely aesthetic grounds, but regrettably aesthetics does not get a vote. More detailed comparisons using cladistics have given ambiguous results, and pareiasaurs and turtles may be unrelated.

Museum specimens like those at Wellwood have received the careful attention necessary to remove the encasing rock and reveal the fine detail of the bones (and no longer with a hammer and chisel as Sidney Rubidge used on that early specimen!). Specimens in a city-based museum, however exciting, are removed from their geological setting. Seeing these specimens in the Karoo makes it far easier to image the world in which they lived.

The Late Permian therapsids include the ancestors of mammals and are often known as the mammal-like reptiles (the latter a name many of my cladistic brethren abhor, but with some resonance). Therapsids include a greater variety of animals than just the pareiasaurs, both carnivores and plant-eaters. Among the

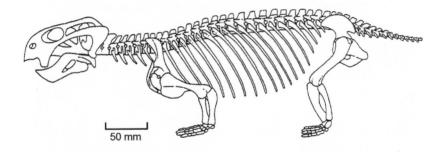

50 mm

Figure 6.5 A reconstruction of the therapsid *Dicynodon*, the vertebrate exemplar of the latest Permian in the Karoo. From Ray and Chinsamy (2003).

plant-eating therapsids are the dicynodonts ("terrible-headed"), with slender bodies, tortoiselike beaks, grinding teeth to shred plants, and broad hips to support the large gut needed for digesting them (figure 6.5). Worldwide, at least thirty-five genera have been discovered in the Late Permian, but only two survived into the earliest Triassic.[2] At some spots in the Karoo the dicynodonts outnumber predators by as much as ten to one, as is the case today for herbivores over predators. The appearance of one dicynodont, *Lystrosaurus*, with its squashed, almost bulldog-shaped face, serves to demarcate the earliest Triassic both in the Karoo and across much of the globe. Some dicynodont fossils have been found curled up, as though entombed in burrows, their long claws suggesting that they may have grubbed for roots. Other dicynodonts also show signs of complex behavior patterns, including probable mole-like forms and browsers. Dicynodonts were not restricted to the Karoo but occurred through southern and eastern Africa as well as Brazil, Antarctica, China, Russia, India, and Scotland.

In Greek mythology the three gorgons were monstrous creatures covered with impenetrable scales, and anyone who set eyes upon them would turn into stone. The gorgonopsids of the Karoo were the major predators of the Karoo, with long, serrated canine teeth and interlocking incisors providing a superficial resemblance to saber-toothed cats (figure 6.6). These large, powerful predators probably shared the gulp-and-go feeding strategy of saber-tooths, but unlike many modern carnivores, gorgons evidently lacked sufficiently massive jaw muscles to rip apart a carcass

Figure 6.6 Gorgonopsid skull. From the collections of the Bernard Price Institute, University of Witswaterand. Photograph courtesy of Chris Sidor.

and so had to content themselves with a few choice steaks. This may hold the secret to the preservation of *Bradysaurus*: the few, poorly developed carnivores had yet to develop the complex adaptations that appear later among the dinosaurs and mammals.

Neither the dicynodonts nor the gorgonopsids gave rise to the mammals. That honor belonged to a different group of therapsids, the therocephalians. They first appear in the Early Permian along with the other therapsid groups, but evolved rapidly into an array of different forms (figure 6.7). By the Late Permian these include some small, probably insect-eating forms with tiny incipient cusps on the back teeth—a preview of a key feature of the later mammals. From these therocephalians the true ancestors of the mammals appear, the cynodonts. The genus *Thrinaxodon* debuts in Lower Triassic rocks and possesses even more mammalian features, including well-developed cusps on the postcanine teeth, and a bony plate separating the nose from the mouth. This secondary palate shifts the breathing passages farther back in the mouth, so an animal can breathe and chew at the same time. The

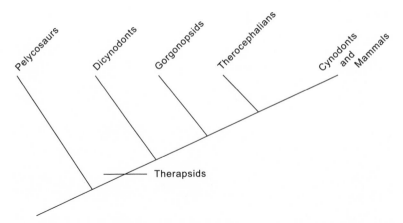

Figure 6.7 General evolutionary relationships of therapsids, gorgonopsids, and cynodonts, which gave rise to mammals.

much higher activity of mammals would not be possible without this development. Although *Thrinaxodon* shows many mammalian characteristics, it is still a long way from being a mammal. Throughout the Early and Middle Triassic, later cynodonts evolved even more mammalian features until the first mammals appear in the latest Triassic or Early Jurassic.

By any reading of the Karoo fossils it is clear the therapsids and their allies suffered massive extinctions, but as with the marine groups, without a good knowledge of the evolutionary relationships of the various vertebrate groups, it is difficult to assess the magnitude of extinction. For example, within a group of reptiles known as the procolophonoids, a strict reading of the fossils suggests that two genera died out near the Permo-Triassic boundary and one line survived: a 67% extinction. In the earliest Triassic at least three new species appear. Not a particularly surprising pattern. Sean Modesto, a Canadian paleontologist who has worked in South Africa with Bruce Rubidge, did a cladistic analysis of the fossils, and this suggests a more complicated history. When the phylogenetic tree is scaled against time (figure 6.8), six lineages were probably present during the *Dicynodon* zone and four made it into the *Lystrosaurus* zone—a 67% survival rate. To reach the merits of this result requires one to accept that several genera now found only in the *Lystrosaurus* zone must have evolved in the latest Permian, but in this case the "ghost lineages" are relatively short

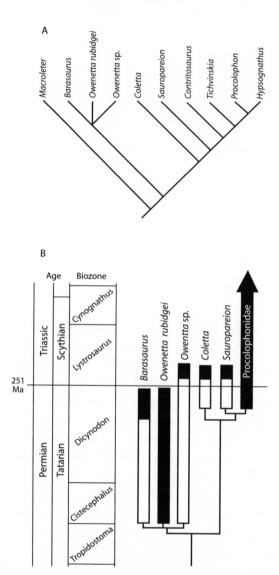

Figure 6.8 Sean Modesto's phylogeny of the procolophonoids. Above is the most likely tree of the group, and below the tree has been rescaled to time. The black bars record the known duration of various genera, with the open boxes the range extensions required by the phylogeny shown above and the known ranges of sister taxa. Because *Owenetta rubidgei* has been found in the *Cistecephalus* zone, the two genera that originated with it are assumed to have evolved at the same time. Abbreviations: *Trop., Tropidostoma; Ciste, Cistecephalus, Dicyn, Dicynodon, Lyst., Lystrosaurus.* Modified from Modesto and others (2001), p. 2050, figures 2 and 3.

and not terribly objectionable.[3] By using this sort of cladistic analysis, we completely change our view of the magnitude of the extinction. Instead of two-thirds of these reptiles disappearing, only one-third did. We do not yet have similar results for therapsids or other groups, but it seems quite likely that the absence of a phylogenetic perspective may have led to considerably overestimating the magnitude of the extinction of all lineages.

· · ·

The rocks of the Karoo have been divided into a series of formations based on the dominant type of rock deposited at a particular period of time. Paralleling this succession of rock types is a series of fossil zones, each named after the most common therapsid fossil found at that time. In 1995 Bruce Rubidge brought together the leading South African workers on Karoo fossils and they put together a consensus sequence of fossil zones for the basin. The *Cistecephalus* zone appears to coincide with the marine Guadalupian, while the *Dicynodon* zone makes up the latest Permian. When Bruce and his colleagues laid the records of fossils throughout the Karoo against these fossil zones they found that nine of twenty-six reptilian genera disappeared at the end of the *Cistecephalus* zone while only seven of the forty-four genera known from the *Dicynodon* zone are found in the Early Triassic (at least one other genus must have survived as a Lazarus lineage because of the reappearance of a group of specialized therapsids in the Middle Triassic) (figure 6.9). Gorgonopsids disappeared entirely, as did the pareiasaurs and most of the therapsids.[4] As Bruce pointed out to me, these data were not collected to assess patterns of extinction (the sort of precise collection within single, well-studied sections described in chapter 3) and this limits the reliability of the results. Fortunately Roger Smith of the South African Museum has worked diligently over the past few years to get better data from the field.

Doornplaats and Lootsberg Pass are two localities close to the Wellwood farm, so after we finished with the museum, Bruce, Sam, Maarten, and I headed off to walk through them. Sam and I were particularly eager to find volcanic ash beds to compare with our

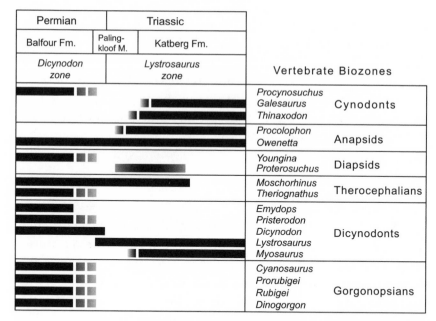

Permian		Triassic		
Balfour Fm.	Paling-kloof M.	Katberg Fm.		
Dicynodon zone		*Lystrosaurus* zone		Vertebrate Biozones
			Procynosuchus *Galesaurus* *Thinaxodon*	Cynodonts
			Procolophon *Owenetta*	Anapsids
			Youngina *Proterosuchus*	Diapsids
			Moschorhinus *Theriognathus*	Therocephalians
			Emydops *Pristerodon* *Dicynodon* *Lystrosaurus* *Myosaurus*	Dicynodonts
			Cyanosaurus *Prorubigei* *Rubigei* *Dinogorgon*	Gorgonopsians

Figure 6.9 Ranges of representative vertebrate groups in the Karoo, with the Permo-Triassic boundary placed at the top of the *Dicynodont* zone. In many cases detailed records of some genera did not exist close to the boundary, so these ranges are shown as dotted lines. The fieldwork of Roger Smith and Peter Ward has subsequently filled in some of these lines. After Rubidge (1995).

work in China. Marine and terrestrial rocks rarely interfinger, making it difficult to precisely correlate a series of marine fossils to their corollaries on land. If we could tie the land and ocean records together, we would be able to tell if both extinctions occurred at the same time. Traditionally, paleontologists have placed the Permo-Triassic boundary between the intervals dominated by *Dicynodon* and *Lystrosaurus*. But in 1967 Nick Hotton, a former vertebrate paleontological colleague of mine at the Smithsonian, demonstrated that the two therapsid genera overlap for about twenty meters across the Karoo. Consequently the exact placement of the boundary was unclear. Should it be at the first *Lystrosaurus*? At the last *Dicynodon*?

Doornplaats has a beautiful farm headquarters along a dry riverbed in the bottom of a broad valley. After saying hello to the farm owners, Bruce and Maarten maneuvered the trucks under some thorn trees to take advantage of the sparse shade; we grabbed our field packs and headed down the riverbed. Several barbed-wire fences later, a broad bend in the river appeared with banks of reddish claystone streaked with white. Bruce halted his forced march and began patrolling the rocks. Head down, hands behind his back clasping a geology hammer that wagged up and down like a tail, we had clearly arrived at something. Many of the white streaks turned out to be bits of bone: the case of a skull here, a long leg bone embedded along the river bank, and then an over-turned skull, the sheared off tusks of a dicynodont staring up like eyes. This is one of the premier fossil localities in the Karoo and it has been preserved in the hope that growing eco-tourism in South Africa will eventually lead to helping South Africans and visitors learn more about the ancient Karoo.

Some paleontologists aren't very good at finding fossils in the field (me), others are outstanding (Bruce Rubidge), and a very few are wizards who magically summon the fossils forward for collection. James Kitching was a wizard, and the highlight of this trip was being in the field with him. Kitch was the son of "Croonie" Kitch-ing, one of the two bothers who had first begun collecting fossils for Robert Broom and Sidney Rubidge in the 1930s. He grew up in Nieu Bethesda, a village near Wellwood, and began collecting fossils with his father at age six. At seven he found his first new species, which Broom named *Youngopsis kitchingi*. More specimens followed and when the Bernard Price Institute for Palaeontological Research, the institution that Bruce now heads, was established in 1945, Kitch became the first employee. By the time he retired fifty-three years later, he had spent almost eighteen cumulative years in the field, most of it tramping up and down the hills and valleys of the Karoo. Driving through the Karoo with him was remarkable, as Kitch recalled a skull from the top of one kopje and a skeleton recovered from a river bottom. With his intimate knowledge of the rocks and fossils of the Karoo, he advised dozens of foreign paleon-tologists on their visits and collected thousands of specimens.[5]

Having had our fill of fossils, our group slowly began making its way back along the river bottom. Bruce remembered some un-usual layers of rock, and Sam and I stopped to collect them in hope that they might be volcanic ashes suitable for age dating.

Ash beds seem to look different everywhere Sam goes, so although there are certain things I have learned to look for (all too slowly from Sam's point of view), we tend to focus on beds that look out of place with respect to the surrounding rocks. At Doornplaats Bruce pointed out a white band along one of the riverbanks, and we also collected several brownish clays almost covered by ledges of the surrounding rock. On the way back to the trucks we discovered a huge (to me, at least) tortoise ambling through a field. He looked at us for a while, and we at him, but the packs were heavy, the sun was hot, and lunch was still several barbed wire fences away.

The next stop was Lootsberg Pass, a classic sequence across the Permo-Triassic boundary alongside a road climbing up through the pass, and providing a spectacular view to the south across the Karoo. We parked the trucks about halfway up the pass, plunged into the brush, and slowly made our way to a small draw where the layers of rock were more completely exposed. Bruce, as usual was wearing shorts and I remain in awe of his ability to wade through a thicket of thornbushes without having his skin shredded. Between brush and erosional scree, it can often be hard to get good exposures of the layers of rock. Since volcanic ash beds are often soft, easily erodible clays, finding good exposures is critical to identifying the ash.

At an earlier stop on this trip we explored a small, several-hundred-meter-high cone of rock where we knew from an earlier report that an ash was present. By pure chance Sam and I happened to climb a gully where the ash was obvious. Five meters to either side and we would have missed it. We often wonder how many ash beds we have walked over, oblivious to their presence. The Permian rocks at Lootsberg Pass are bluish-gray to olive-green mudstones, capped by fine sandstones. We did not know the exact position of the Permo-Triassic boundary, but Bruce knew it was close to the first appearance of maroon-red mudstones. If there are any ash beds at Lootsberg Pass, they eluded us that day, and have continued to elude us on subsequent visits as well. Despite our best efforts, and those of Bruce and his colleagues, we still have not found any ash beds close to the PT boundary in the Karoo. We know that volcanic ash beds were deposited near the PT boundary in Antarctica, and we are sure that eventually someone will find them in the Karoo.

Doornplaats and Lootsberg Pass have figured prominently in other studies of the end-Permian mass extinction in the years since my first visit. Peter Ward, a well-known paleontologist at the University of Washington who played a significant role in studies of the Cretaceous-Tertiary mass extinction, teamed up with Roger Smith of the South African Museum to launch detailed studies of the vertebrate extinction patterns. Roger is one of the best field sedimentologists in the world, as well as an outstanding contributor to understanding of the fossils of the Karoo. I spent a week in the Karoo with him in July 1998 and witnessed his almost mesmerizing ability to pull up to an outcrop along the side of a road and conjure up sand bars migrating down long-vanished rivers, floods inundating the plains with silt, and a family of *Dicynodon* drowning in a burrow.

As they began scouring the Karoo for new data on the pattern of extinction, Peter and Roger were concerned with the same issue that had brought Sam and me to South Africa: how to tie the land and sea records together. They chose to focus on the carbon isotope record instead of ash beds. The Karoo lacks the limestones so useful for carbon isotopes in marine rocks, so Ken MacLeod, a paleontologist and geochemist now at the University of Missouri and former student of Peter Ward's, found fossil soil horizons with soil nodules and dicynodont tusks. Tusks are hard and dense, so often preserve a good isotopic record of the plants the dicynodonts ate. They found a large shift in carbon isotopes beginning with the first *Lystrosaurus* and reaching a peak almost coincident with the last *Dicynodon* (at the top of the overlap between *Dicynodon* and *Lystrosaurus*). Above this spike the isotopic values reverse. The number of samples they could collect was far fewer than in marine sections, so it is impossible to say whether there were multiple shifts in the carbon cycle. This remarkable isotopic shift is similar to that at Meishan and other marine Permo-Triassic boundary rocks. In 2005 a more detailed isotopic study was published by Peter and colleagues (this time I was one of the many coauthors) and strongly suggests that the overlap zone coincides with the marine Permo-Triassic boundary. Since the carbon shift occurs over several meters, this suggests that while the extinction was rapid it was not instantaneous.[6]

Doornplaats covers thousands of acres, stretching from the river bottom we explored, well up the side of the surrounding hills. The several thousand meters of rock exposed extend from the *Cistecephalus* fossil zone up through the *Dicynodon* fossil zone almost to the Permo-Triassic boundary. This great vertical exposure of rock poses one of the great problems with older fossil collections. Often the only locality data is "Farm Doornplaats." All well and good if the farm is flat and lies within the same fossil zone, but when several zones are present on a single farm, older collection data are almost useless for establishing the fossil zones.

The truth is that as scientists develop new questions, previously collected fossils are often of little use. The declining utility of older data is self-evident in fields such as high-energy physics or molecular biology but often seems to be heresy to museum scientists who think that their collections can address any issue. Nothing could be further from the truth. Fossils are essential for documenting patterns of geographic distribution, providing specimens for the essential work of establishing evolutionary relationships between groups, and as sources for other studies. (Who would have guessed that comparing California condor eggs from the 1930s and 1970s would show that the pesticide DDT had so thinned condor eggs that mothers were crushing the young as they incubated the eggs?) Each new generation must make new collections to address new issues, ensuring that museum collections continue to expand.

In the Karoo the vast collections of Bain, Geddes, the Kitchings, and Rubidge lacked the precise stratigraphic data needed to establish the pattern of extinction. At Lootsberg Pass, Bethulie, and other Permo-Triassic boundary spots around the Karoo, Roger Smith and Peter Ward spent weeks in the field making new collections and noting the precise stratigraphic position of fossils. During several years of effort in the field, Roger and Peter logged the distribution of eight Permian species, of *Lystrosaurus* crossing the boundary, and another seven Triassic species. Roger and Peter showed that the shift in isotopes occurred in a finely laminated mudstone bed between the Permian olive-gray mudstones,

and the overlying Triassic maroon mudstones and large sandstone beds. This is just the horizon where *Dicynodon,* the gorgonopsid *Rubidgea,* and the other Permian vertebrates disappear,
leaving *Lystrosaurus* to continue on into the Triassic, while both
the pattern of extinctions and the discovery of the finely laminated beds at the apparent boundary suggest that the vertebrate
extinctions were rapid and happened at about the same time
across the Karoo.

I have emphasized the story from the Karoo because it was one
of the first areas studied, and because I know the area well. But
as Roger and Peter were studying the Karoo vertebrates, Michael
Benton, a vertebrate paleontologist at the University of Bristol in
England began working with colleagues in Russia on rocks from
the southern Urals of Russia. The vertebrate fossils here are second only to those of the Karoo in abundance and quality and, like
the Karoo, have attracted the attention of generations of paleontologists. The rocks record a sequence of mudflats, and river channels, much like South Africa, representing rivers and lake ecosystems. Pareiasaurs, gorgonopsids, and dicynodonts were all present
as well. Some 82% of the vertebrate families became extinct at
the Permo-Triassic boundary. This is similar to earlier turnovers
between stratigraphic units, but the big difference was that the
number of genera and families remained low in the Early Triassic,
instead of expanding as had happened previously.[7]

· · ·

Dicyodon, Bradysaurus, and the other Permian animals of the Karoo
lived in an environment lusher and wetter than the modern Karoo
desert. Broad, meandering rivers wandered slowly across the landscape, spilling over the levees during floods and depositing sandbars in the river bends. Carcasses and bones carried down the river
often beached in these sandbars. Well-articulated fossils are often
found entombed in the muds that spilled across the levees during
floods, and these muds formed the greenish and maroon mudstones straddling the Permo-Triassic boundary.

As the mountains of the Cape Fold Belt grew during the Permian amalgamation of Gondwana, they behaved as any adolescent

mountain chain and rapidly shed sands, conglomerates, and other debris into the surrounding lowlands. High mountains change climate patterns as well, with rain and snow increasing on the windward side of the mountains and a rain shadow developing in the lee. In the Sierra Nevada Mountains of California the comparatively lush western foothills contrast with the barren desert of Nevada and Utah. Similarly in South Africa, as the Cape Fold Belt grew it cast a rain shadow across the Karoo, drying out what had been a lush semitropical environment. As the mountains grew, erosion increased and more rock, sand, and gravel was dumped into the rivers, changing the form of the rivers. Sandstones become thicker and more complex; mudstones became less frequent. Meandering rivers were displaced by a network of anastomozing channels, known to geologists as braided river systems. Braided rivers develop when there is so much sediment in a river that the channels frequently choke up and force the flow into a different course. The channels are generally less sinuous than in meandering rivers. Four different factors can produce a change from meandering to braided rivers: an increase in the slope of the river, an increase in the amount of sediment, a decrease in the amount of water flowing through the river, or an increase in the variability of rainfall. Or all four together.

The onset of braided rivers in the Karoo Basin has long been viewed as a gradual change, with the growth of the Cape Fold Belt decreasing rainfall patterns and increasing erosion rates, producing rivers choked with sediment. If we could study this transition at a fine enough time scale, we should see the braided rivers develop first in the south, adjacent to the mountains, and then spread slowly to the northwest. Moreover, the switch from meandering to braided rivers should have begun just after the uplift of the mountains. Dating the uplift of mountain belts is easier than dating sediments and the mountain-building episodes in the Cape Fold Belt occurred about 258 and 247 million years ago. The Permo-Triassic boundary dates to 251 million years ago and this posed a real problem if climate was driving the change in sediment style. Roger and Peter's careful fieldwork showed that the transition from meandering to braided rivers happens very close to the last *Dicynodon* and the first *Lystrosaurus* all across the Karoo. In

2000 they argued that braided rivers developed across the Karoo just at the Permo-Triassic boundary, reflecting not the rise of the Cape Fold Belt Mountains, but the extinction itself.[8] The change in sedimentation style corresponded with the end of *Dicynodon* and the appearance of *Lystrosaurus* in too many sections, and the sharp shift in carbon isotopes coincided at Lootsberg Pass and Bethulie. With plants no longer holding back sediment, sand, mud and gravel would clog the channels, and this, rather than mountain building, was what Peter and Roger believe caused the change in river systems. Their argument makes intuitive sense to me. Growing up in southern California, I well remember the rush to replant devastated hillsides after a bad season of brush fires in the hope (usually forlorn) that the grasses would take hold before the winter rains triggered mudslides and the creeks became torrents of mud and boulders.

Even if this shift occurred simultaneously across the Karoo, persuasive evidence of a regional change is no indication of a global pattern, and indeed one of the easiest traps for a paleontologist is to grandly extrapolate from hard-won field data. Peter and Roger realized that one of the best tests of their argument was to look at Permo-Triassic sections on land in other parts of the world, and sure enough, braided rivers appear near the boundary in Australia, Europe, and elsewhere. Some of the best examples of this shift come from research that was being done at the same time Peter and Roger were working in South Africa.

At the southern end of the Ural Mountains, just north of the border with Kazakhstan, thick conglomerates of sand, gravel, and rock develop at the Permo-Triassic boundary. Russian geologists had long related this change to increased sediment eroded off the young Ural Mountains, a remarkably similar idea to the Cape Fold Belt and the Karoo. Recent fieldwork suggests a pattern strikingly reminiscent of the Karoo: a sharp increase in erosion with sand and gravel choking the rivers and changing patterns of sedimentation, probably due to destruction of most plants, and a much warmer, drier climate. As with the Karoo, there is little evidence of active mountain building that could have provided the sediment.[9]

Ideas are often contagious and some philosophers of science have begun studying the patterns of infection among scientists much as epidemiologists study disease transmission. A new idea, a new way of looking at the world spurs other geologists to view their field areas in a new way. The papers by Ward and Smith led geologists at the Indian Institute of Technology at Kharagpur to rethink the Raniganj Basin northwest of Calcutta, and they report the onset of semi-arid conditions, an increase in erosion, and a change in sediment style coinciding with the Permo-Triassic boundary.[10]

While the Karoo, Russian, and Indian sections all suggest something happened at the Permo-Triassic boundary to increase erosion and change sedimentation patterns, many have inferred that plant abundances dropped catastrophically to allow the increased erosion. But there are few studies of plants to show whether their abundance declined as predicted. One problem is that a drop in abundance does not necessarily translate into a loss of species, and as Andy Knoll pointed out two decades ago, plants may more easily weather catastrophic losses of biodiversity without going extinct. Plants are better preserved and better studied in other regions, providing a test of these models.

Ancient rivers in eastern Australia, in an area known as the Bowen Basin, are more revealing of the relationship between plants and changing rivers. The Bowen Basin stretches north of Sydney toward Townsville, in northern Queensland. Broad, sinuous rivers rimmed with levees deposited Late Permian sediments. Large lakes lay between the rivers with peat swamps in the lowlands. With time, the buried peats eventually developed into widespread coal beds. The abrupt change in the carbon cycle seen in marine sections (chapter 7) is also found here, providing a critical marker for the Permo-Triassic boundary. Only a meter above the isotopic shift, coals disappeared and sheets of sand invaded the basin, probably deposited by flash floods in broad, straighter rivers. Geologists often call these sheet flood deposits. As in South Africa and Russia, they reveal a rapid shift from Permian environments. Permian rocks have abundant plant fossils, and even aside from the coals, the muds deposited during floods contain abundant organic debris. All this is missing from the Triassic, again suggesting that plants have disappeared. How sudden this change was, however, is a bit unclear, for in the northern part of the Bowen

Basin the pollen record begins to record a gradual change in the plants well before the Permo-Triassic boundary.[11]

Coals disappeared across the globe at the Permo-Triassic boundary, not just in the Bowen Basin. In Antarctica, elsewhere in Australia, India, Russia, and China, coals suddenly end near the Permo-Triassic boundary. Scattered, thin coal seams have been found in the Middle Triassic, but thicker coal seams generally do not reappear until the end of the Middle Triassic or the Late Triassic.[12] Coals develop in peat bogs, where acidic waters favor the slow, steady accumulation of plant material. After the peat is buried, time, heat, and pressure eventually turn it into coal. The acidic environment of a peat bog is one to which relatively few plants have adapted. Wipe out these plants, and these lowland settings will be relatively uninhabited until a new suite of plants acquires the necessary adaptations. Peat bogs also need moist, humid conditions, and sufficient room for long-term accumulation of decaying plant matter. Removing any of these will prevent the development of peats, and thus coals.

In the Sydney Basin of eastern Australia, the coals disappear at the same time as a pronounced shift in plant fossils, which Greg Retallack believes corresponds with the Permo-Triassic boundary. In many parts of the southern supercontinent of Gondwana one of the most common plant fossils is that of the long, tapering leaves from the *Glossopteris* tree (figure 6.10). These disappear in the Sydney Basin with the last coal, and are followed by a shrubby plant known as *Dicroidium* that was well suited to dry environments. In Australia only four plant genera seem to have survived this extinction, and the number of plant species in the early Triassic was very low for several million years.[13]

The simplest explanation for the disappearance of the coals is the long-term change in climate associated with the northward drift of Pangaea during the Permian. This gradually brought more continents close to the equator, ending the long Carboniferous to Early Permian Ice Age and producing a warmer and drier global climate. The ferns, treelike lycopsids, tree ferns, and other odd plants that produced the widespread coal swamp forests of the Carboniferous and Early Permian were ill suited to these warmer

Figure 6.10 The characteristic Early Triassic *Dicroidium* (top) and the Permian tree *Glossopteris* (bottom).

and drier climates. The plants of the cool, wet Carboniferous were replaced by new groups of plants that constructed the Mesozoic fields and forests. Analogous to Jack Sepkoski's three great Evolutionary faunas, the older plants are known as the Paleophytic flora, and their replacements the Mesophytic flora. The new groups include many conifers, ginkgoes, cycads, and ferns. While conifers remain widespread, ginkgoes and cycads are mostly found in the Jurassic Park section of botanical gardens, as all these groups have almost disappeared under the metastasizing post-Cretaceous profusion of flowering plants.

This Paleophytic to Mesophytic floral transition has been well known to paleobotanists for decades. In 1984, as part of a magisterial overview of plant evolution Andy Knoll argued that plants were little affected by the end-Permian mass extinction. He proposed that extensive seed banks in the soil, rhizomes, and other patterns of plant growth naturally make plants more resistant to mass extinction than animals. In Andy's view, the change of floras was a gradual event over tens of millions of years, and the end-Permian mass extinction had no discernible impact on plants.

Discernible. There's the rub. Studying the fossil record is similar to a zoom lens: narrow studies stand out in sharp focus and reveal interesting details, but a wide-angle view is needed to see if the pattern is general. This is a paleontological analog of the Heisenberg uncertainty principle (one can measure the position of an electron, or its momentum, but not both at the same time). In paleontology, one can develop either a global view of changing patterns of diversity over millions of years (at least during the Paleozoic) or the sharp detail of a single locality, but the latter at the cost of not really knowing whether this spot is typical of the world. Global studies of plant diversity, even when broken down into continent-sized regions, show little evidence of a mass extinction among plants, in sharp contrast to the pattern seen among marine animals.[14]

The rocks in the Bowen Basin seem typical of other regions, where coals disappear with the Permo-Triassic boundary. But some geologists have argued that if coals were disappearing gradually through the Permian and Triassic as the climate changed, pure chance would dictate that some of these changes would occur near the Permo-Triassic boundary. Undue emphasis on these few areas might blind us to a longer, gradual conversion.

At the Permo-Triassic boundary in East Antarctica coals drop out at the boundary. The Permian sediments alternate between calm, sinuous rivers across a broad flood plain and higher-energy, braided fluvial channels, but in the Triassic, braided rivers predominate and the climate shifts from consistently wet to a seasonal

dry. Overall, this is very similar to the Karoo. Stephen McLoughlin
of the University of Melbourne in Australia and his colleagues
compared this pattern to six other regions: Brazil, Nevada, the
Karoo, India, Australia, and elsewhere in Antarctica (figure 6.11).
McLoughlin argues that the data reveals a gradual disappearance
of coal and an equally gradual appearance of the red beds that
are so common in the Triassic. For example, red beds first appear
in Nevada in the Lower Permian, but this is the only one of the six
regions where red beds are apparent before the Late Permian.[15]
Despite their claims, when I look at the data I see a pattern equally
consistent with a sudden change in climate at the Permo-Triassic
boundary.

In some sense there may not be enough data to definitively resolve whether climate changed suddenly at the Permo-Triassic
boundary or not, but the trend toward rapid change is consistent
through recent studies. My own view is that plants suffered some
degree of disruption at the close of the Permian, but it was not
of the magnitude we see among marine invertebrates, or even
among the animals of the Karoo. Dicynodonts in the Karoo, coals
in Australia, and the proliferation of braided rivers all point toward widespread changes across many continents at the boundary.

Insects are often overlooked as fossils, and rarely have a starring
turn in accounts of mass extinctions. In the Late Permian, however, they provide further evidence of considerable disruption of
terrestrial ecosystems. There are thirty-seven described orders
within the Class Insecta (orders are the next rank below class), and
twenty-two orders are known from the Permian. The only mass
extinction insects experienced occurred during the Permo-Triassic transition, with about eight orders disappearing and another
five losing many families (figure 6.12). By comparison, only one
or two insect orders have disappeared since the Permian. My colleague at the Smithsonian, Conrad Labandeira, has compiled data
on the distribution of fossil insect families through their history
in the same fashion that Jack Sepkoski (one of Conrad's advisors
in graduate school) did for marine animals. Fossil insects show a
truly catastrophic extinction, with family diversity plummeting
from about sixty in the Lower Permian to near zero at the Permo-
Triassic boundary. The insect fossil record remains sufficiently

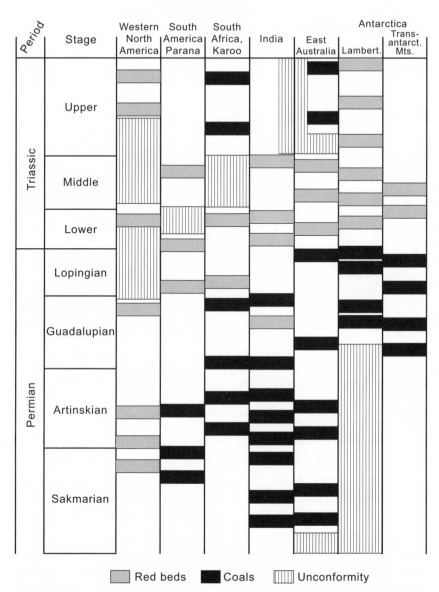

Figure 6.11 Coals disappeared in the Late Permian, and in many areas were replaced with red sandstones ("red beds") formed in arid and often warm environments. Although some changes began following the end of the Permo-Carboniferous glaciation in the Lower Permian, change seems to have accelerated in the Late Permian. Redrawn after McLouglin et al. (1997). Copyright Antarctic Science LTD, 2005, reproduced from *Antarctic Science* 9:293.

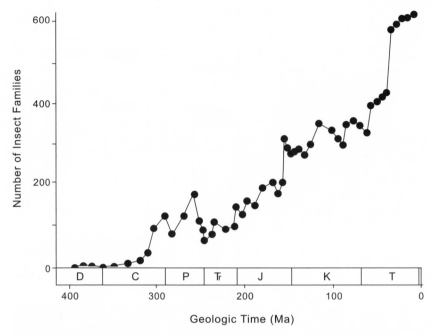

Figure 6.12 Diversity of fossil insect families from their origin during the Devonian to the present, showing the sharp drop in diversity during the end-Permian mass extinctions. Reprinted from Labandeira and Sepkoski (1993). Copyright AAAS.

poor that Conrad is unable to say how many of these extinctions occurred at the boundary. Since many orders persisted, the extinction was not complete, but there was clearly a significant turnover in the insects at this time.[16] Several of the orders that disappeared were among the Palaeoptera, a primitive group of insects that includes dragonflies and others that cannot fold their wings back over the body upon landing, and leave them jutting out to the sides. Beetles, grasshoppers, and cockroaches all have a simple hinge for the wing, allowing these modern neopterous insects to fold the wings back against the body (figure 6.13). Neopterous insects could gain access to many new habitats denied the paleopterous forms, and today about 98% of modern insects are neopterous. Most neopterous insects also separate the larval stage from the adult by a resting stage or pupa, allowing larvae and adult to have different ways of making a living. This adaptation also helps insects survive in highly seasonal environments, and Conrad Labandeira has suggested that as more seasonal environments

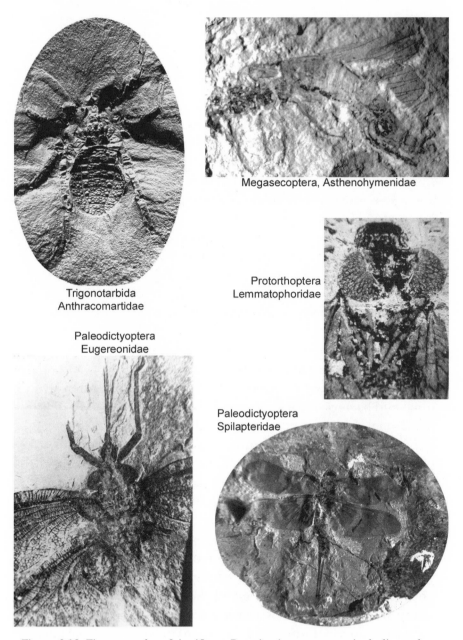

Trigonotarbida
Anthracomartidae

Megasecoptera, Asthenohymenidae

Protorthoptera
Lemmatophoridae

Paleodictyoptera
Eugereonidae

Paleodictyoptera
Spilapteridae

Figure 6.13 Five examples of significant Permian insect groups, including paleo-dictyopterids. Photos courtesy of Conrad Labandeira.

spread during the Permian this encouraged the spread of insects with a pupal stage.

. . .

I originally titled this chapter Revenge of the Fungi, and although I discarded this for reasons that will be clear shortly, we should celebrate the contributions of these unheralded elements of terrestrial ecosystems. Our consideration of this closest relative to animals is normally limited to their gastronomic possibilities. Fungi are common on rotting logs because they are essential decomposers and are particularly important in degrading trees and other woody plant remains. Heather Wilson, one of the leading students of millipedes and other multilegged arthropoden wonders, recently pointed out to me that the animals found today in leaf litter from a forest are almost the same as those from 450 million years ago when plants and arthropods first invaded the land. Other than ants and termites, the advent of forests has produced few new animal decomposers because in some sense other decomposers remain oblivious to the existence of the forest. Until fungi and bacteria have produced a nicely rotten log, there is little the animal decomposers can accomplish.

This is by way of introduction to one of the more peculiar aspects of the Permo-Triassic boundary: the remarkable abundance of apparent fungal remains. This "fungal spike" is reminiscent of a burst of fern spores immediately after the Cretaceous-Tertiary boundary, and as with the fern spike, this transient burst of fossils may indicate a decimation of terrestrial ecosystems, with the fungi growing on decaying plant matter just as the early Tertiary ferns were weedy opportunists taking advantage of the catastrophe. This pattern is the reverse of the pattern with other plants where there is less evidence of catastrophe. What appear to be fungal remains include both long chains of cells representing the hyphae of fungi, as well as fungal spores, usually called *Reduviasporonites* (figure 6.14). Henk Visscher of the Laboratory of Paleobotany and Palynology at Utrecht University and his coworkers Henk Brinkhuis and Cindy Looy, and Yoram Eshet of the

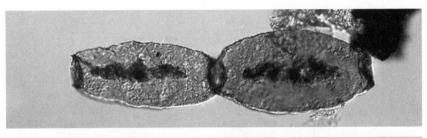

Figure 6.14 The fungal spores *Reduviasporites*, provided by Clinton Foster and used with permission, after Foster et al. (2002). Photographs taken by Mike Stephenson.

Geological Survey of Israel, have been diligently tracking this problem for almost a decade.[17]

Fungal remains are almost nonexistent in rocks below the extinction but increase to almost 100% of all pollen, spores, and similar remains at the marine extinction peak in Israel and the southern Alps. In South Africa the abundance of *Reduviasporonites* increases to 100% at the floral extinction boundary. The spike is found in both marine and terrestrial rocks, but this is not particularly unusual since such remains from land are easily carried to the oceans by both wind and rivers. The ubiquity of these distinctive cells has been viewed as providing a time line allowing correlation between many different boundary sections, although in Italy, Germany, Greenland, Svalbard, Siberia, Austria, Australia, Turkey, Madagascar, India, Kenya, and China, the increase in fungal remains occurs in rocks below the marine extinction (figure 6.15). The Meishan Permo-Triassic locality described in chapter 3 has been particularly useful because the radiometric dates tell us the increase in fungal remains began at least 500,000 years before the marine extinction.

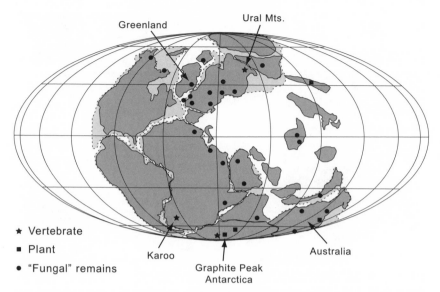

Figure 6.15 Permo-Triassic terrestrial localities with the continents in their Late Permian positions, showing localities of terrestrial vertebrates, plants, and reports of the "fungal" spike.

Roger Smith's work in the Karoo suggests that the Permo-Triassic transition lies between the greenish mudstones of the Upper Permian Balfour Formation and the laminated maroon mudstones just below the Katberg Sandstone. A one-meter section of green and red siltstones and sandstones just below the Katberg have abundant fungal spores and some woody material.[18]

Fungi are such prolific decomposers of trees and shrubs that the spike must reflect the catastrophic decay of forests, and indeed of the whole terrestrial ecosystem. When Henk Visscher and his group studied the fungal spike in the southern Alps and Israel, they also noticed that pollen from gymnosperms declined at the same point, suggesting a large-scale loss of plant biomass. Since there was no scent of an impact when they published their work in 1996, they suggested that vast amounts of carbon dioxide from the Siberian flood basalts caused the disruption on land.

The geographic breadth and magnitude of the fungal spike appears to be compelling evidence for a widespread and catastrophic destruction of terrestrial ecosystems at the Permo-Triassic boundary. Just as important, the onset of the fungal spike at Meishan suggests that at least some terrestrial events may have begun 500,000 years before the extinction in the oceans. This last bit is hardly consistent with mass extinction due to an extraterrestrial object. Game over?

Naively assuming that all was well in the world of pollen and spores (and not wanting to delve too deeply, I must admit), I had assumed the case for the fungal spike was unambiguous. But one of the joys of working at the Smithsonian Institution is the serendipitous arrival of visitors. One day in March 2000 I had an email from Robert E. Lee (I am not making this up) announcing that he would be in town in a few days, and would like to talk with me about the extinction. Upon arrival Bob Lee turned out to be a delightfully unrepentant hippie masquerading as an algal specialist at Colorado State University. Bob quickly got to the point, and much to my surprise suggested that the "fungi" were remarkably similar to various algae. He made a pretty good argument, and in a few weeks followed up his visit with a long letter and several pages of images comparing various Permian fossils with living algae. Since I had known Henk Visscher for several years and knew he was highly respected within the paleobotanical world, I was initially suspicious of Bob Lee's claims. The fungal spike seemed such a wonderful marker of destruction of terrestrial habitats that I must admit, Bob seemed to be ruining a great story, which also annoyed me. If the "fungi" were largely algae then there was no indication of terrestrial destruction, but a bloom of pond scum. But I quickly discovered from Andy Knoll and others that Bob was very respected as an algologist and the author of a well-regarded textbook.

Others had noted the same problem. In 2001 Sergey Afonin from the Paleontological Institute in Moscow drew attention to the remarkable similarities between *Tympanicysta* (which Visscher had pointed out should be called *Reduviasporonites*) and a group of green algae known as the Zygnematales. Although identification of the close living relatives of some fossils can be relatively unambiguous, some identifications are much harder. Fossils with relatively simple shapes can be hard to identify, and if no living relatives exist, the problem is even more difficult. Afonin studied beautifully preserved earliest Triassic material from the Vologda region of Russia where the shape of the cells, the connections between them, and some other details are very similar to green algae. Indeed these fossil cells are so well preserved that what appears to be chloroplasts are preserved.[19]

Support for the views of Bob Lee and Sergey Afonin came from long and patient work by Clinton Foster of Australia in a paper that is almost Victorian in understatement.[20] Many recent papers on the fungal spike had been trumpeted broadly to the scientific community in prestigious journals like *Proceedings of the National Academy of Sciences* and *Geology*. Foster published in *Palynology*, properly treasured by specialists on pollen and spores but all but ignored by even other paleontologists. Foster's group properly began with a long and detailed taxonomic description of the genus *Reduviasporonites*, with paragraphs devoted to cell shape, the outer cell wall, the inner body, cell material (Afonin's cholorplast), folding patterns of the cell wall, the terminal rim where cells connect, and the formation of chains of cells. This is the real meat of paleontology and his taxonomic study is based on samples from many different parts of the world, spanning the Middle Permian (Capitanian) into the Early Triassic. (In fairness to Visscher and others, they never claimed that *Reduviasporonites* was confined to the Permo-Triassic boundary, only that it became unusually abundant in that zone.) Almost as an afterthought, Foster's group appends a geochemical analysis to the end of the paper, but this is the gem that delivers the coup de grâce to claims that fungi blossomed on the rotting corpses of end-Permian forests.

As discussed in more detail in the next chapter, carbon isotopes can reveal what organisms were feeding on. If *Reduviasporonites* was a fungus living on decaying plants, the carbon isotopic values should be only slightly different from the vegetation, as is the case today. Instead specimens of *Reduviasporonites* have carbon isotope ratios much more negative (lighter) than most trees and other woody plants. The difference between *Reduviasporonites* and the woody tissue reported by Clinton Foster seems much too large for these specimens to be fungal. Moreover, *Reduviasporonites* contains chemical fossils in the form of the organic carbon compounds preserved in cell walls. These chemical fossils are most similar to distinctive products from algae; missing were any of the distinctive chemical fossils that characterize fungi. Although it may seem beyond belief that such characteristic fossils could be preserved in rocks 251 million years old, similar structures have been recovered from rocks more than 2 billion years old.

If Clinton Foster is correct and this burst of cells truly is green algae rather than fungi, it dramatically changes our interpretation of the Permo-Triassic boundary and the earliest Triassic. Instead of the acid-rain-induced destruction, their reinterpretation as green algae could suggest that there was change in the environment that allowed the green algae to bloom like weeds in shallow aquatic environments.

· · ·

The end of the gorgons, the paleodictyopterid insects, and many plants establishes the spread of the extinction on land as well as sea, but did they happen at the same time as events in the oceans? If the event was largely marine, then extinction scenarios focused on the oceans would suffice as an explanation. Marine anoxia, changes in sea level, and bursts of methane gas must fail as sole causes because their effects are largely limited to the oceans. An extraterrestrial impact or similar catastrophic cause requires extinctions to happen at the same time (or as close to the same time as we will be able to see in the rock record). In an ideal world Sam Bowring and I would have found a nice volcanic ash bed in Lootsberg Pass, just up the road from Wellwood in the Karoo. In the absence of ash beds to directly date and tie to our work in south China, more indirect methods are required. In the Karoo the only means to correlate between land and sea are the now discredited algal spike and changes in carbon isotopes, and these provide only suggestive evidence that the terrestrial and marine events coincided. Far better to search for a spot where both marine and terrestrial fossils are found together.

Richard Twitchett, a young paleontologist now at the University of Plymouth has found such a spot. His graduate work at the University of Leeds with Paul Wignall focused on the Early Triassic recovery, and he has visited Permo-Triassic boundary sites around the world. Tall and gangly, Richard is a patient geologist who has already published some exceptional results. Working along the eastern coast of Greenland, he revisited the Schuchert Dal region where a thick sequence of marine muds and silts contains both great marine fossils and terrestrial pollen and spores. The Permo-Triassic boundary lies near a change from green and gray silt-stones to dark gray muddy siltstones; the carbon isotopes drop rapidly over about three meters of sediment and many fossils disappear. This carbon isotope shift is the best indicator of the position of the boundary: the first diagnostic Triassic conodonts, the microfossil standard for biostratigraphy, do not appear until about sixteen meters higher in the section where fossils again become abundant. Land must have been reasonably close by, for along with ammonoids, brachiopods, and conodonts, pollen grains are abundant. The distinctive Late Permian pollen begins to decline before the apparent boundary.[21]

Twitchett's fossils suggest that animals in the oceans and plants on land began to disappear at the same time, and before the carbon spike. Assuming the carbon shift here and at Meishan is the same, and the conodonts appear in both places at the same time (the first is a reasonably safe assumption, the second a bit more dubious), Twitchett used the absolute ages from Meishan to calculate the rate of sedimentation in Greenland. This suggested that the collapse of the marine ecosystem occurred in perhaps as little as 20,000 to 30,000 years. The shift in carbon after the apparent extinction could indicate the shift is a *consequence* of the extinction rather than reflecting the *cause* of the event, but unlike our study in Meishan, Twitchett and his group have not yet collected enough data for statistical analysis. In the absence of the statistical approach I discussed in chapter 3, it maybe premature to read too much into the timing of the carbon isotope shift.

Returning to the animals of the Karoo, we can now be fairly confident that they disappeared at the same time as the marine extinction. Their disappearance, along with the end of so many insect groups, changes in plants, and the shift in rivers from slow, meandering streams to the rapid sedimentation of braided streams demands that the cause of the extinction affected both land and sea. We turn now to a final set of clues as to the causes of the extinction, including the meaning of the shift in carbon isotopes and other changes in chemical cycles.

As I write this on a rare sunny summer day in 2003, our rainfall is fifteen inches above normal. The daily deluge should continue for the coming week and I am more worried about my garden drowning than desiccating, quite the reverse of our normal August worries in Washington. Europe has been suffering through paralyzing heat, with temperatures soaring above 100°F in places, and the French President is accusing his countrymen of abandoning the old and infirm in their rush to the beaches. Ignoring the difference between weather and climate, some reporters have been quick to link these odd weather patterns to the growing levels of carbon dioxide in the atmosphere. By now we can all recite the implications of pumping millions of tons of carbon dioxide from fossil fuels into the atmosphere: carbon dioxide is a greenhouse gas, providing an insulating blanket to the Earth's atmosphere. As the amount of carbon dioxide increases in the atmosphere, temperatures climb and seasonal fluctuations become stronger.

But this is only part of the entire cycle of carbon. I have already noted that major changes in the carbon cycle coincided with the end-Permian mass extinction, and there is emerging evidence for an equally profound shift during the end-Guadalupian mass extinction as well. This raises a number of important questions. Does the change in the carbon cycle reflect the cause of the extinction, or was it the result of the extinction? Photosynthesis partitions carbon on the Earth into two great reservoirs: the organic carbon reservoir of almost everything living as well as of coal, oil, and other organic remains, and a larger reservoir of inorganic carbon that has not passed through living things, including the limestone in a reef and carbon dioxide in the atmosphere. At the close of the Permian a massive amount of organic carbon was released into the atmosphere and oceans. Where it came from is the critical issue. Among the possibilities is that it reflects the carbon in all the plants and animals that died, or the burning of massive coal deposits in Siberia, or the release of methane gas trapped in sediments on the outer shelf of the continents.

Sorting out these issues requires a deeper understanding of the isotopes of carbon, how they flow through the oceans and atmosphere, and how knowledge of the isotopes of carbon tells us about the shifts of carbon between the organic and inorganic reservoirs. View this discussion of shifting carbon isotopes as a view into the health of the planet, in the same way a doctor's tests tell him or her about your health. Here the doctors are geochemists using mass spectrometers to reconstruct the history of their patient hundreds of millions of years ago.

Much of the carbon cycle involves shifts between the two great reservoirs of carbon, one of organic carbon built of almost everything living and all buried organic matter, and a second inorganic or carbonate reservoir of the remaining carbon, from carbon deep within the earth's mantle to limestone in tropical reefs. The oceans and atmosphere act as the fulcrum shifting carbon between these two reservoirs via interchange between the oceans, atmosphere, and soils. Organisms, living and dead, are the major component of this short-term cycle, but the cycling of carbon through organisms, the oceans, and the atmosphere is sufficiently quick that we can safely ignore it to focus on the longer-term, geological cycle.

The long-term cycle involves interchange between carbonate rocks like limestones, buried organics such as coal and oil, the absorption of carbon dioxide from the atmosphere by weathering of rocks, and the release of carbon dioxide from deep in the earth via volcanoes and through plate tectonics. Within the oceans, carbon can be removed as organic particles that fall to the sea bottom and become buried, thus increasing the organic carbon reservoir, or as particles of carbonate produced on reefs or in other places that enlarge the second, carbonate reservoir. Organic carbon can also be buried on land, in coal swamps, peat bogs, or in lakes. Carbon is added to the carbonate reservoir from the deep earth through volcanoes and the spreading of the oceanic plates along the global midocean ridge system. Carbonate carbon is removed from the atmosphere via weathering of feldspars and other calcium and magnesium silicate rocks in mountain ranges. Weathering generates calcium and carbonate ions that are carried by rivers to the sea. Burning fossil fuels completes the processes, mixing organic carbon back into the carbonate pool. The same thing happens over longer time spans when coals are eroded or oil seeps to the surface along faults. In both cases slower weathering of the oil or carbon returns them to the carbonate pool.

Explaining how the organic and inorganic carbon reservoirs are created or how we can track the shifts between them requires a brief detour into carbon isotopes and their behavior. Remember that the different isotopes of an element have the same number of protons, but differ in the number of neutrons. Each isotope is still carbon however many neutrons it has. The radioactive isotope of carbon (C-14) permits dating of bones, burnt corncobs, and other archeological debris over the past 100,000 years but decays rapidly so we do not need to consider it further. The two other isotopes, C-13 and C-12, do not experience radioactive decay. The critical feature for all that follows is that during photosynthesis, plants and microbes prefer to utilize carbon dioxide made of C-12. It may seem odd that one measly neutron can make such a difference, but it is enough to change the efficiency of photosynthesis. Thus photosynthesizing organisms have more C-12 in them than the average ratio between the two elements. Consequently everything that eats photosynthetic organisms is also enriched in C-12, with a slight additional enrichment as the carbon moves up the food chain. Coal, oil, peat, the black goo at the bottom of a swamp, because they are all ultimately derived from products of photosynthesis as is all buried organic material, are likewise enriched in C-12.

Just as organic particles record enrichment of C-12 from photosynthesis, nonorganic sedimentary particles likewise record the ratio of C-12 to C-13 at the time they were formed. Now if carbon simply cycled between the organic and inorganic reservoirs, the ratio of C-13 to C-12 would never change. But imagine that a vast coal bed forms. This removes a large volume of carbon, enriched in C-12, shifting the ratio between the two isotopes remaining in the oceans and atmosphere. Conversely when the coal is eroded, or burned, organic carbon with more C-12 is returned to the oceans and atmosphere, and the C-13 to C-12 ratio shifts in the other direction. In fact, any event that adds more organic carbon to the oceans, sequesters more inorganic carbon, or otherwise alters the amount in the two reservoirs will be recorded as a shift in the C-13 to C-12 ratio in the rocks and fossils formed at that time.

Following the carbon requires a meter to tell us whether carbon is flowing in or out of the organic carbon reservoir. Geochemists have developed just such a meter. Known as $\delta^{13}C$, this is the difference between the ratios of the two isotopes in a sample relative to a known standard, and is traditionally reported in parts per thousand, or $^0/_{00}$. A zero value simply means the carbon ratio is the same as in the standard (a fossil cephalopod from the Cretaceous Pee Dee Formation and hence known as the Pee Dee belemnite). Geologists around the world refer to the same standard so analyses made in Nanjing, Cambridge, and Krakow can be compared directly.

A shift of $\delta^{13}C$ to more negative values indicates that some volume of organic carbon has been added to the oceans and atmosphere. Similarly, a shift from negative to positive (or less negative) normally means that inorganic carbonate has been added, perhaps from weathering of rocks. Leaving rapid shifts aside for the moment, longer-term values of $\delta^{13}C$ are also revealing about the state of the carbon cycle. A high $\delta^{13}C$ value (above zero) is produced by high rates of organic carbon burial as coal or as organic-rich muds, while a low value (below zero) indicates extensive carbonate burial often of a large amount of limestone.

Thus far I have treated the carbonate and organic reservoirs as if they were single entities, but in fact each is composed of many smaller pools that differ strongly in their $\delta^{13}C$ values. Today, within the inorganic reservoir, carbonate rocks have a value of about $1^0/\text{oo}$, the surface ocean about $2^0/\text{oo}$, and carbon dioxide venting from a volcano has about $-5^0/\text{oo}$. Organisms differ in the extent to which they prefer C-12 to C-13, and this difference in what geochemists call isotopic fractionation produces characteristic differences in the $\delta^{13}C$ ratio of different organisms. The average carbon isotope value of soils and plants on land is $-25^0/\text{oo}$, while the average in the oceans is only $-20^0/\text{oo}$. Microbes that produce methane, so-called methanogens, prefer C-12 to C-13 much more than do other organisms, so the methane trapped in arctic permafrost or in deep sea sediments has a very light carbon ratio of at least $-65^0/\text{oo}$.

These different carbon pools also vary greatly in size. There are about 82 million gigatons of carbon in carbonate rocks such as limestone (a gigaton is 1 billion metric tons of carbon), 14 gigatons of carbon in organic rocks (coal, oil, peat), and at most some 10,000–15,000 gigatons in methane locked in deep-sea sediments. The total biomass of terrestrial plants and animals is at most 1,000 gigatons with all life in the oceans contributing perhaps 3–5 gigatons of carbon. Why are these differences worth mentioning? It is the combination of the $\delta^{13}C$ value and the volume of a particular pool that determines how significant the pool may be in changing the overall $\delta^{13}C$.

Mass extinction may seem the most obvious way of shifting $\delta^{13}C$. Kill off most plants and animals, a goodly number of microbes as well, and sequester the carbon out of the system by burial on land or in the sea, and the carbon isotopic ratio of the whole ocean will shift. But not enough. Remember that there are perhaps 1,000 gigatons of carbon in all living plants, animals, and microbes. In the Permian the value was probably even lower. If we add 1,000 gigatons of carbon at, say $-25^0/\text{oo}$ $\delta^{13}C$ to the oceans, the volume of carbon in the oceans is so great that there will be little change in the global value of $\delta^{13}C$. Fortunately there are many other sources of carbon to trigger the shift in carbon isotopes associated with the end-Permian mass extinction, but before we consider these possibilities, let us take a closer look at the actual pattern of change in carbon isotopes at the Permo-Triassic boundary.

. . .

At Meishan the carbon isotopic record in the latest Permian is relatively stable between +3–4^0/oo through the Changhsingian (figure 7.1), indicating a relatively high level of deposition of organic carbon. The isotopic ratio begins a gradual decline at bed 23, then drops sharply at the Permo-Triassic boundary to values as low as −2^0/oo, then quickly recovers to between +1 and 0^0/oo through the earliest Triassic. Jonathan Payne, a graduate student at Harvard, has compiled one of the first detailed records of Early Triassic carbon isotope values, revealing at least four significant Early Triassic swings in the carbon cycle before the carbon cycle settles back down to a value of about 2^0/oo in the Middle Triassic. The Meishan record tells us that two things happened virtually simultaneously: the addition of sufficient organic carbon to the oceans and atmosphere to cause the $\delta^{13}C$ value to drop to about −2^0/oo, *and* a more permanent change in how much organic carbon was being buried during the earliest Triassic.[1] This same pattern of a very abrupt shift coincident with the extinction horizon and the longer-term shift of the $\delta^{13}C$ value shows up in many other Permo-Triassic boundary sections around the world.

In 1986 a drill rig much like those used to search for oil was helicoptered up the Gartnerkofel in the western Alps to punch a core through the Permo-Triassic boundary. Known to geologists as Gartnerkofel-1, or GK-1 (it is hard to think there will ever be a GK-2, but geologists are ever optimistic), the core preserves a detailed record from the last bit of the Permian through the Early Triassic. Sedimentation was slow when the rocks of the Gartnerkofel were deposited, even slower than at Meishan, and there are few gaps in the record. A team of nineteen geologists studied the core in detail, examining the fossils layer by layer, making a wide range of chemical measurements and searching for evidence of an extraterrestrial impact.[2] The pattern of change in carbon isotopes is the same as at Meishan, dropping from about +3^0/oo to −1^0/oo before stabilizing in the earliest Triassic at just above

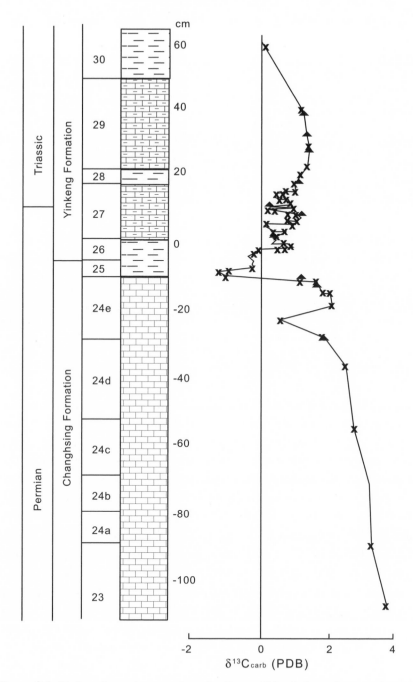

Figure 7.1 Carbon isotopic record at the Permo-Triassic boundary section at Meishan, China. This data is from Jin et al. (2000).

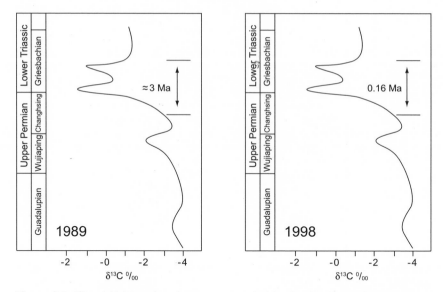

Figure 7.2 The shift in carbon isotopes recorded in the GK-1 core from western Austria, with the original estimates of the length of the change in carbon isotopes from 1989 of about 3 million years, based on the number of conodont biostratigraphic zones. With the radiometric dates from Meishan, it now looks like the shift happened in about 160,000 years.

$1^o/oo$ (figure 7.2). Unlike Meishan, however, the GK-1 core has an interesting double dip, a second negative spike in the earliest Triassic. Other localities across the Permo-Triassic boundary from Greece east through Iran, India, and China were studied at the same time. Some have gaps near the boundary, and others had slow rates of sedimentation, but overall they show a similar pattern to that at Meishan (figure 7.3).[3] In the past decade many other areas have been sampled—from New Zealand, Australia, Japan, China, Pakistan, Iran, Armenia, Europe, and north to Spitsbergen, Greenland, and as far west as British Columbia. All show the same basic pattern and are correlated with the conodont biostratigraphy. This sharp drop in the carbon isotope ratio is so pronounced, and so well studied, that it has served as a marker of the Permo-Triassic boundary in sections where fossils are absent. As we will see later, multiple shifts in the carbon cycle have recently been recorded from the earliest Triassic in Australia, China, and elsewhere. This suggests caution in using carbon isotopes alone to identify the exact position of the Permo-Triassic boundary.

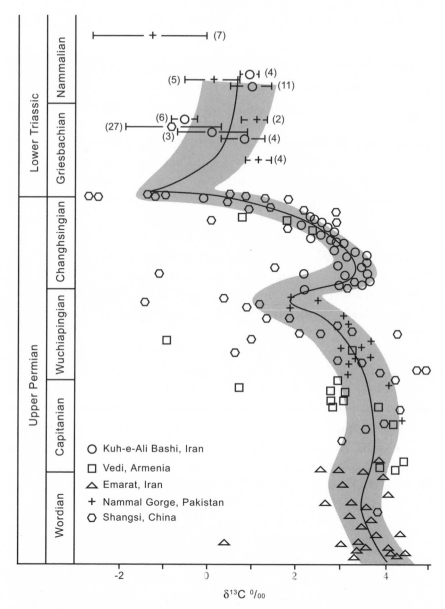

Figure 7.3 The pattern of isotopic shifts looks very similar in localities around the world. Redrawn from Amon Baud et al. (1989).

Up to this point I have been describing measurements of changes in the carbonate reservoir, $\delta^{13}C_{carbonate}$. The isotopic ratio can also be measured in preserved organic matter as $\delta^{13}C_{organic}$. Since isolating organic material in the laboratory is more challenging, analyses of the organic record have lagged studies of the shift in the carbonate reservoir but that gap has closed and many recent studies include measurements of both the change in the carbonates and the organics. The most striking result of these newer analyses is the magnitude of the shift in the organic carbon. As I mentioned above, average values of $\delta^{13}C$ for marine plankton are about $-20^0/00$, $-15^0/00$ for marine plants and algae, and -20 to $-30^0/00$ for the organic carbon buried in sediments. Late Permian records of marine organic carbon have values of about -24 to $-26^0/00$. But at the Permo-Triassic boundary, the $\delta^{13}C_{organic}$ values plunge to $-33^0/00$, or even $-38^0/00$ in some places. A huge shift. One confounding factor is that the fractionation between C-12 and C-13 increases as the amount of carbon dioxide in the atmosphere increases. As we will see, such an increase is highly likely, so at least some of the shift in the $\delta^{13}C_{organic}$ values could be explained by this increase in carbon dioxide. But the shift may also require the addition of a good deal of very negative organic carbon, much lighter than the average values seen for marine plants and algae.[4]

The shifts in $\delta^{13}C_{organic}$ and $\delta^{13}C_{carbonate}$ indicate that a huge volume of organic carbon was added to the oceans and atmosphere at the Permo-Triassic boundary. Now we need to establish which of the various pools of organic carbon could produce this shift. But before we do so, there are a few additional issues to be resolved. First, most of the areas I have discussed thus far lay along the margin of the supercontinent of Pangaea, or within large continental basins open to the ocean, as in China. Could the shift reflect conditions only adjacent to continental masses, or does it reflect changes across the global ocean? If the carbon shift encompasses the whole ocean, this greatly enlarges the amount of carbon that was involved. Second, did the shift in isotopes occur in both the organic and carbonate reservoirs at the same time? If they did not change in parallel, we will have to consider different explanations, such as a role for carbon dioxide from the Siberian volcanism. Finally we also need to convince ourselves that the record is accurate, and has not been altered in the past 250 million years.

Taking these issues in order, we will begin with whether the shift was global. Establishing the geographic extent of an isotopic shift is fairly easy in rocks from the past 100 million years: oceanic islands and sunken seamounts contain a good record for geologists to sample. Any Permian islands have been swept up by continental drift and mashed against the continents. Fortunately one of the benefits of the plate tectonic revolution has been intensive global work to untangle the wreckage of such collisions. The detritus of Permian seamounts have been identified in Japan and British Columbia; each area records the same carbon isotopic shift seen elsewhere. Because the opening of the Atlantic since the Triassic has steadily reduced the size of the Pacific Ocean, Japan and British Columbia were much farther apart in the Permian than they are today. So these records confirm that the isotopic shift affected the global ocean, and not just areas adjacent to the continents.

Moreover, the organic and carbonate carbon $\delta^{13}C$ changed in unison in sections in both Japan and British Columbia, as well as in the GK-1 core in Austria and at Meishan. From this we can conclude that the carbonate and organic reservoirs were both enriched in light carbon across the Permo-Triassic boundary (the evidence is more equivocal in some sections, however). There is also some curious evidence that the magnitude of the difference between the organic and carbonate records may differ with latitude, although the evidence is far from compelling.[5]

Maarten de Wit, the same geologist at University of Capetown responsible for my first visit to the Karoo, has uncovered multiple negative shifts in the organic carbon ratio in six different basins from South Africa, Madagascar, and India. These areas cover a vast area today, but during the Late Permian they all lay near 60° south latitude in the southern portion of Gondwana. Martin and his colleagues discovered multiple sharp shifts of −5 to −15⁰/00 $\delta^{13}C_{organic}$ from the latest Permian into the earliest Triassic. Each negative shift is separated by strong positive values of $\delta^{13}C$, suggesting repeated additions of very light organic carbon.[6] Sedimentation rates at Meishan were so low that it is not surprising we did not find multiple excursions there. If records from other regions substantiate multiple isotopic shifts, this raises the possibility that the extinction occurred over an appreciable, if still brief, interval of time, a finding with significant implications for identifying the cause of the extinction.

Finally, can we be sure that the record is accurate? Recovery of such a similar pattern in many different places does give us confidence that the pattern is real. Some negative isotopic values may be artifacts, caused by heat and other effects of burial that can alter isotopic records, but this plagues the organic carbon record far more than the carbonate record. A potential problem with the $\delta^{13}C_{organic}$ record may be important. Recall that I said that different plant species might fractionate organic carbon to different degrees and thus their remains would have different isotopic values. Replacement of one plant by another across a landscape would thus change the $\delta^{13}C$ of carbon feeding into an ocean, and change the overall organic carbon isotopic record without any real change in the carbon cycle.[7] Analyses have now proceeded to the point where the real magicians of this work, such as Roger Summons of MIT, can isolate specific organic molecules from the organics preserved within the rock, and this is just what is needed to determine whether the source of the organics has changed through the course of an extinction, although this still occurs infrequently with Permo-Triassic studies. Analysis of the Hovea-3 drill core in the Perth basin of western Australia revealed that the shift was triggered by the disappearance of organics from organisms feeding on several levels of the food chain (heterotrophs) and their replacement by plants and algae.

. . .

Let me summarize what we know at the moment. A massive volume of light, organic carbon was introduced into the oceans and atmosphere coincident with the mass extinction. Dates from the ash beds at Meishan in China suggest that this spike happened in less than 160,000 years, far less than the 3 million years estimated in 1989. This brief, transient excursion was accompanied by a longer-term change in the carbon cycle toward less burial of organic material than in the Late Permian. Records of this carbon shift from the Alps, China, and oceanic islands now part of Japan reveal that both the carbonate and organic carbon reservoirs shifted in parallel, and that the shift was global. Although the organic carbon record is plagued by potential alternations, there are suggestions of multiple introductions of organic carbon and the hint of more carbon introduced in high latitudes, particularly high southern latitudes, than near the equator. Now we can turn to investigate the possible sources of the organic carbon.

First, can we estimate just what we mean by "a large volume of organic carbon was introduced at the boundary"? In other words, how much carbon? When I reached this point in my work on the extinction during the early 1990s I was stumped on how to proceed, and in particular how to work out what processes could produce the shift in the carbon cycle. But I stumbled upon a wonderful paper by two European geochemists that solved my problems. Alejandro Spitzy and Egon Degens of the Universität Hamburg described two models for understanding carbon isotope fluctuations. Since the process was simple enough for me to work out on a computer spreadsheet, I was soon happily testing different scenarios.

We know the size of the carbon shift at the Permo-Triassic boundary (from $+3-4^0/00$, down to $-2^0/00$, for the carbonate reservoir), the values of the various different organic pools that could have contributed, and that the shift may have influenced the global ocean. The equations from Spitzy and Degans allowed me to take this data and determine the volume of each pool that would be required to produce the observed shift in the carbon cycle. In other words, if we know that the carbon isotopic ratio shifted by, say $4^0/00$, how much volcanic carbon dioxide, at $-5^0/00$, $\delta^{13}C_{organic}$ marine organic carbon at $-20^0/00$, or methane, at about $-65^0/00$, is required to force this shift, and is it available? Comparing the need against the likely size of each pool will help evaluate the probability of each potential source.

I discovered that some 40,000 gigatons of carbon from carbon dioxide vented by volcanoes is needed, 6,500–8,500 gigatons of organic carbon, either marine or terrestrial, and only about 2,500 gigatons of methane. The Late Cretaceous Deccan flood basalts in India probably released about 1,000 gigatons of carbon dioxide, and while the Siberian flood basalts are much larger, 40,000 gigatons seems wholly unrealistic. Moreover, remember that we are considering only the inorganic carbon change, but volcanic carbon would be unable to produce the very negative shifts seen in the organic carbon record. None of this means that Siberian volcanism was not involved in the extinction, or that it could not have induced the carbon shift indirectly, only that the carbon shift was not directly a result of carbon dioxide released during the eruptions.

Organic carbon is quite abundant, particularly on the continental shelves, and oxidation of this material could produce the observed shift. But such marine organics are generally oxidized only when sea level drops and the organics are exposed to the atmosphere. With little evidence for a drop in sea level across the end-Permian boundary, this scenario seems unlikely. Oxidation of organic material on land is more plausible, but estimates of the total carbon of all the peat, plant detritus, and humus in soils is only about 1,000 gigatons, far less than the 6,500 or so gigatons needed to drive this change in carbon isotopes.

Far more promising is methane, although I say this with great trepidation. Methane is highly enriched in C-12 (meaning it has a negative value), so less of it is required to cause a significant shift in the $\delta^{13}C$ value. In 1993 I was quite enthusiastic about invoking methane. Only 2,500 gigatons are needed to produce the observed P-T shift, well below the estimated 10,000 gigatons of methane then estimated to be locked into sediments on the outer continental shelves. Moreover, a burst of methane is the easiest explanation of the very light records of −38 or −33⁰/₀₀ in organic $\delta^{13}C$.

Initial estimates of the volume of sea-floor methane made it the solution to all problems. At an annual meeting of the American Geophysical Union a few years ago it did not seem to matter much what the scientific problem was, the answer was methane. Carbon shifts? Methane! Sudden greenhouse warming? Methane! Giant landslides found in the Atlantic and Mediterranean? Methane released by earthquakes! I began to feel I had wandered back to the age of Dr. Fussbudget's Elixir to Cure All Ills. Despite the vague whiff of snake oil, there is no denying that for decades geologists had been blithely ignorant of the amounts of methane buried in ocean sediments.

In the ocean, methane is found as an icelike solid, often with considerable water, and so is known as methane hydrate, as mentioned previously. Formed during decay of organic mater by methane-producing bacteria, gas hydrates are most common in the outer continental shelves at depths of 300–500 meters where the cold and the pressure from the overlying water keeps it locked into the sediments. Methane hydrates are particularly common in high latitudes where the cold allows methane to persist in shallower waters. (Methane also occurs in peat swamps and a fair bit of methane can form in tundra). As oil-drilling rigs penetrated deeper waters, they would occasionally keel over when the drilling hit a layer of methane, as the solid bottom of the sea floor melted away. Having ones' drilling platform tip over was hardly popular among oil companies, or their insurers, and this stimulated efforts to define the extent and size of these methane fields. Collapsing drilling platforms is not the only reason for interest in methane. It is a far more powerful greenhouse gas than carbon dioxide, and though it only persists in the atmosphere for 10–50 years until it is converted to carbon dioxide, while it lasts it can have a dramatic impact on climate. If methane bleeds slowly into the atmosphere over decades or centuries, the impact is greatly magnified. There may be 5,000 gigatons of carbon in the coal, oil, and natural gas reserves, but until recently, estimates were as high as 10,000–16,000 gigatons of methane carbon.

These estimates were only tenuously based on actual measurements of methane. Methane leaks out of drill cores as they are recovered from the sea bed, so there are few good measurements. Gerald Dickens of Rice University and colleagues measured methane abundance in three drill holes in the Blake Ridge off Georgia where methane-rich zones corresponded with an acoustic echo mirroring the sea bottom as seen on seismic profiles of the continental margins. Such bottom-simulating reflectors have been interpreted as marking the bottom of methane gas deposits. Assuming that most bottom-simulating reflectors indicated methane trapped below the sea floor, geologists could estimate the volume of sediment that contained gas, multiply by the volumes of gas at Blake Ridge, and come up with really large numbers.

It would be nice to know that the bottom-simulating reflectors did indicate methane, so in 2003, Alexi Milkov of the Woods Hole Oceanographic Institution and his colleagues (including Gerald Dickens) sampled methane concentrations above and below the bottom-simulating reflectors at Hydrate Ridge off Oregon. They found high volumes of methane in one small area but the amount of methane was much lower than at Blake Ridge, and so they concluded that the global amount of gas hydrate might be only 2,000 to 3,000 gigatons instead of 16,000 gigatons.[8]

Extrapolating from these results to the Permian is difficult. With the earlier estimates of methane volume, the 2,000 or so gigatons of carbon needed to trigger the isotopic change at the Permo-Triassic boundary were within reach in the deep sea, but not readily available. Although several million years is needed to buildup a significant subsea supply of methane, it can be released quickly. Methane hydrates rapidly become unstable if the temperature rises; pressure falls through a drop in sea level or through a sudden shock like a massive earthquake. Sea level was rising during the end-Permian mass extinction, so there is little chance that a drop in sea level released a large volume of methane. With rising sea level, the only way to release methane would be by warming permafrost as seawater floods shallow marine areas, or by heating the oceans. The new estimates of gas hydrate abundance do not change the conditions for methane release, but do limit the plausibility of this argument.

The enormous Siberian coal deposits underlying Siberian flood basalts could rescue the methane hypothesis. Vaporizing coal deposits is an excellent source of carbon dioxide and thus global warming. New work on early Eocene rocks from the Norwegian Sea suggests that the rapid global warming at this time may have been due to the conversion of carbon-rich sediments to methane by volcanic activity. Heating carbon-rich sediments to 100–200°C will at least partially convert the carbon to methane. In the Norwegian Sea 50 to 200 times more carbon would have been released by conversion of carbon-rich sediments to methane than from simple release of volcanic carbon dioxide.[9] We will return to the methane problem in the following chapter.

Up to this point I have assumed that the changes in the carbon cycle affected the whole ocean from the surface to the deep sea. This assumption may not be correct. If our radiometric dates from Meishan accurately reflect the rapidity of the extinction, the organic carbon release occurred in just a few hundred thousand years or less. Although shifts of carbon between the ocean and atmosphere happen quickly, the carbonate and organic carbon reservoirs respond much more slowly. Consequently, the dates from Meishan demand that the carbon excursion during the extinction probably did not occur in deeper water and was largely restricted to the surface waters of the ocean. Since the volume of the surface waters (say to a depth of 100 meters) is far smaller than the volume of the whole ocean, less carbon is required to drive the shift than in the estimates above, which assumed that the whole ocean was involved. If this is true, we might be able to explain the shift in carbon as the elimination of primary productivity during the mass extinction, which will cause carbon isotope values to gradually drift toward about $-6^0/00$, and by a drop in the amount of carbon buried (figure 7.4).

So explaining the carbon shift turns out to be fairly complicated. To return to the questions at the beginning of the chapter, the critical issue is whether the shift in carbon isotopes occurred coincident with the mass extinction or just afterward. If the latter, then the shift may reflect the effect of the extinction rather than the cause. Distinguishing between these two alternatives requires very high temporal resolution, however. Most geologists have argued that the isotopic shift reflects the cause of the extinction and have interpreted it in that light. The rapidity of the extinction as measured in Meishan, and Richard Twitchett's suggestive work in Greenland that the mass extinction on land and sea occurred before the beginning of the shift in carbon isotopes (chapter 6), each suggests that the carbon shift reflected the effects of the extinction rather than the cause. I pointed out that the fossil record from Greenland is too poor for geologists to have great confidence that the extinctions and isotopic shift were truly diachro-

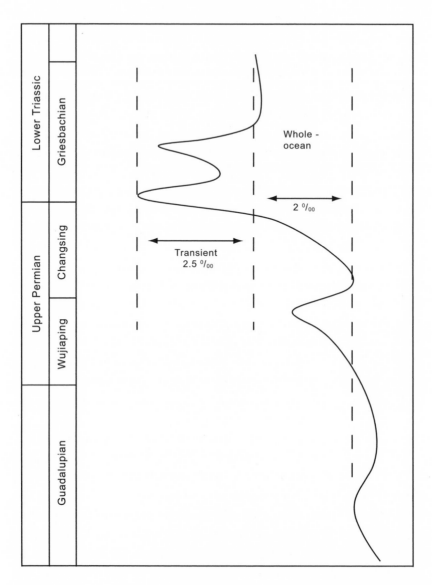

δ¹³C °/₀₀

Figure 7.4 The carbon isotopic shift has been interpreted as representing a change in the isotopic values of the entire ocean. The alternative shown here was originally suggested by the late Bill Holser of the University of Oregon, with the most negative spikes only occurring in shallow waters due to the disappearance of photosynthesis during the extinction. The longer term shift from +4 to about +1 is due to a smaller rate of burial of organic carbon.

nous, but Richard could be right. We have seen that release of methane is difficult to reconcile with independent evidence that sea level was rising rather than falling, so unless we can develop a different means of releasing methane from the continental shelves, this source of methane may fail. But the conversion of extensive Siberian cola beds to methane by volcanic activity may have been sufficient to release massive volumes of methane. The significant shift in organic carbon does track the shift in carbonate carbon, but we now know that this could be affected by an increase in carbon dioxide in the atmosphere or by a relative increase in the proportion of plants and other photosynthetic organisms contributing to the organic carbon pool.

We are left with three possible explanations for the carbon shifts at the Permo-Triassic boundary. First, release of methane via volcanic heating of the Siberian coal beds. Second, much of the shift may not be a whole-ocean change in the carbon isotope ratio, but a surface-water phenomenon reflecting a decline or elimination of photosynthetic activity in the oceans. Finally, there is a considerable reservoir of organic carbon in terrestrial soils, and increased weathering and erosion of this material after the landscape had been denuded during the extinction might have contributed to the isotopic shift.[10]

· · ·

Carbon is not the only element to cycle through the ocean and atmosphere, or the only one to record dramatic changes at the Permo-Triassic boundary. Sulfur and strontium have different cycles than carbon, and thus reflect other changes in the physical environment. Examining the sulfur cycle is particularly important because increases in sulfur in the atmosphere can produce sulfuric acid and thus acid rain.

Sulfur in the ocean generally combines with oxygen to form sulfate, but certain bacteria can change sulfate to sulfide, which in organic-rich, oxygen-poor muds is converted to pyrite. Burial of muds effectively removes sulfide from the ocean. In contrast, sulfate is moved out of ocean water during formation of evaporite deposits such as gypsum and anhydrite. There are two isotopes of sulfur, S-32 and S-34, and the formation of evaporates faithfully preserves the ratio of S-34 to S-32 (known as $\delta^{34}S$) in the oceans as they were formed. Just as sugar dissolves in hot water, sulfur dissolves in magma deep within the Earth. But there is a limit to the amount of sugar that one quart of water can hold; the remainder lies swirling around the bottom of the pot. In the same way, the volume of sulfur dissolved in magma is limited, but the 1991 eruption of Mount Pinatubo in the Philippines revealed that volcanoes have options unavailable to ordinary cooks. Twenty times more sulfur came out of Mount Pinatubo than geologists expected given the volume of magma erupted. Like a vision from Dante's hell, Mount Pinatubo evidently had a layer of bubbling sulfur near the top of the magma chamber. As the volcano erupted, the sulfur was carried high into the atmosphere, cooling summer temperatures in the northern hemisphere by up to 2°C and warming winter temperatures by 3°C for several years. In the Karoo Basin, rocks at the Permo-Triassic boundary are rich in sulfide. The most likely explanation is the introduction of enormous volumes of sulfate into the atmosphere that bacteria then converted to sulfide. In the absence of any evidence for a natural increase in continental weathering, a plausible explanation is that weathering increased due to acid rain generated either by a bolide, or by extensive volcanic eruptions.[11] I also wonder whether the increased sulfur might explain the extensive sedimentary pyrite that Paul Wignall and others have documented from just above the Permo-Triassic boundary.

Just as carbon isotopes record the flow of carbon between the organic and carbonate pools, fluctuations in the strontium ratio across the Permo-Triassic boundary provide an index of changes in global tectonic activity. The shifting balance between the two isotopes of strontium, Sr-86 and Sr-87, reflects changes in sources of the two isotopes. Strontium-86 is released by basaltic magma, particularly from the global rift system that drives the drift of the continents. As the rate of sea-floor spreading increases, more water will flow through the young basalt, releasing strontium and lowering the ratio. Weathering of granitic rocks shifts the strontium ratio in the opposite direction. So lighter values reflect more mantle-derived material; heavier values mean an increase in continental weathering.[12]

Strontium ratios change dramatically though the Late Permian and Early Triassic. The magnitude of this shift is greater than at almost any other point in the last 600 million years. A significant drop in the strontium ratio in the uppermost part of bed 24 at Meishan is coincident with the mass extinction. The strontium ratio then reverses, climbing suddenly to a much higher value in bed 27. These changes have to be seen against a longer-term shift in strontium isotopes. The strontium ratio reached a very low value in the Guadalupian just before the first mass extinction and then began climbing, signaling greater input from continental weathering. Recall from chapter 6 widespread red sandstones become more abundant at this time as Pangaea moved into more equatorial latitudes. With increased weathering in equatorial latitudes, nothing more may be needed to explain this long-term increase in the ratio. The straightforward explanation for the reversal at the Permo-Triassic boundary is a quick pulse of mantle strontium in just a few hundred thousand years interrupting a long interval of greater continental weathering.

Kunio Kaiho from Tohoku University in Japan collected some of the sulfur and strontium data from Meishan, and also discovered very tiny iron-silica-nickel particles at the top of bed 24 at Meishan.[13] The particles, in Kaiho's view, condensed out of the hot vapor of an impact and are similar to grains found at other impact horizons, including the Cretaceous-Tertiary boundary. Kaiho has proposed that the strontium and sulfur isotopic data and the iron-silica-nickel particles can be explained by a massive extraterrestrial impact that in turn triggered an enormous release of volcanic material, including a large volume of sulfur. The mass extinction was caused by a combination of the effects of impact, follow-on climatic change, and acid rain produced from the sulfur in the atmosphere. The sudden input of mantle-derived strontium comes from the Siberian flood basalts. Other geologists, including me, believe the sulfur and strontium data can be explained by Siberian volcanism alone and do not require an impact. The Japanese group's evidence for impact rests on the nickel-rich particles and the earlier claims of extraterrestrial fullerenes by Luann Becker's group.

Leaving discussion of Kaiho's hypothesis that the strontium and sulfur isotopes suggest impact-induced volcanism for the next chapter, I want to return to the sulfur isotopes and an alternative explanation that they reveal evidence of low oxygen levels in the deep ocean. To do so requires a brief excursion into deep-sea sediments. The limestones and sandstones so typical of shallow oceans are missing from the deep sea where siliceous cherts and clays replace them. The cherts are almost entirely formed by the rain of tiny microfossils known as radiolarians, whose skeletons are made of silica. As the skeletons accumulate in the deep ocean they form layered chert deposits. Hematite, an iron oxide (basically rust) is also common in the deep sea, so deep-sea cherts are often a deep, burgundy red. Remember that subduction of oceanic plates has destroyed deep-sea sediments from the Permian, but a few precious bits were scraped onto the sides of Japan and elsewhere. From these slivers of the deep sea we find that in latest Permian and Early Triassic rocks the hematite disappears and is replaced by pyrite, an iron sulfide mineral that forms in the absence of oxygen. Along with the change from hematite to pyrite there is also an increase in the $\delta^{34}S$ ratio. The sulfur ratio was low though the Middle Permian but increases near the Permo-Triassic boundary, signaling the onset of anoxic conditions. The sulfur isotopes then reverse just at the boundary before returning again to a higher level and presumably an anoxic deep ocean.

Today oxygen is abundant in the deep sea, delivered by water masses cascading down from shallow waters, but in restricted basins such as some lakes or the Black Sea, where there is little or no exchange with the open ocean, the deep waters can become chemically quite distinct from the surface waters as the decay of organic material removes oxygen. Such an anoxic deep ocean may seem very bizarre, but there is no reason why such oceans cannot exist and there is growing evidence that they may have been quite common for much of Earth history. The most plausible interpretation of the changing sulfur ratios is that a stagnant, anoxic ocean developed in the deep ocean essentially separated from the surface ocean through the same stratification we see today in the Black Sea. This stratified ocean began in the Permian and persisted into the Early Triassic, with a brief interval of oxygenated

deep water just at the boundary. The same shift to anoxic sediments has been reported from British Columbia as well as Japan,
and since the Pacific Ocean was even broader during the Permian,
Yukio Isozaki used this evidence to propose his "Superanoxia" hypothesis that the entire ocean was stratified with a deep anoxic
layer for perhaps as long as 20 million years from the Late Permian
into the Middle Triassic.

A group of microbes known as the green-sulfur bacteria thrive
in such environments where they convert sulfate to sulfide, leaving
a trace of distinctive organic products in the sediment. Isolation
of these organic compounds, known to the trade as biomarkers,
is pretty unequivocal evidence for the presence of this noxious
brew of low-oxygen and also sulfur-rich waters. Evidence from
both the Hovea-3 well in Australia and Meishan reveals that these
conditions must have been fairly widespread across the Permo-
Triassic boundary and into the earliest Triassic. Another group
has also found evidence for sulfidic deep waters during the latest
Permian in Greenland.[14]

The focus of this chapter is on the changing chemistry of the
oceans and atmospheres during the Permo-Triassic transition and
what this tells us about the extinction. The sulfur isotopes have
raised the issue of deep-sea anoxia, and for the sake of complete-
ness, it seems appropriate to continue with a discussion of the evidence for low-oxygen conditions in shallow water.

Paul Wignall and Tony Hallam found evidence for anoxia in
shallow-water sediments across the boundary in Pakistan, the Dolomites, south China, western United States, and the island of Spitsbergen in the high Arctic. They discovered black shales, pyrite,
and fossil assemblages dominated by a few brachiopods and bivalves such as *Claraia*. These fossils are suspected of preferring low
oxygen conditions. Their analysis of the rocks revealed that sea
level was rising and Wignall and Hallam suggested that the increase in anoxic sediments followed the rising seas. In other
words, mass extinction was due to a rising sea level moving a layer
of anoxic water through shallow marine waters. As their research
has progressed over the past decade, they have marshaled additional evidence to support the anoxia hypothesis, including
changes in bioturbation of the sediment that can be a sensitive

indicator of environmental conditions. Suggestions of shallow-water anoxia have been found in low-latitude areas that were part of Pangaea (Dolomites and Pakistan), regions separated from Pangaea (south China), as well as high-latitude regions such as Spitsbergen, indicating the anoxia was globally distributed.[15]

Other geologic evidence is less favorable to the shallow-water anoxia model and probably the best place to see this evidence is in the high Arctic islands of Canada. Benoit Beauchamp of the Geological Survey of Canada is a compact, energetic geologist who spends much of his summers in the deep, steep-sided valleys cut into Arctic islands when sea level was much higher. These exposed fjords reveal the magnificently exposed Carboniferous, Permian, and Triassic rocks of the ancient Sverdrup Basin. The changing rock types show that warm, tropical conditions dominated during the Early Permian, with widespread reefs, green algae, and benthic foraminifera. By the Middle Permian, a cooler-water assemblage of bryozoans, echinoderms, brachiopods, and siliceous sponges replaced the tropical species. As siliceous sponges came to dominate the basin in the Late Permian, their spicules formed extensive chert deposits. Every indication is that the cherts formed in a very cold environment. These massive silica factories were not limited to the Sverdrup basin, but extend along the northwest margin of Pangaea as far south as Nevada. The latest Permian destruction of this extensive, very cold, silica-rich assemblage, what Beauchamp calls "silica factories," occurred as rising sea levels drowned the basin. Cherts are missing from the basin until the Middle Triassic.[16]

Recall that Pangaea moved northward during the Permian. This could provide a simple explanation: the change of fossils represents the movement of the Sverdrup Basin from tropical into more Arctic settings. But the studies of the magnetic field preserved in the rocks show the basin moved only 10–15° of latitude during the mid to Late Permian, far too little to have produced a transition from tropical to polar climates.

Formation of such extensive chert deposits demands an adequate and relatively continuous supply of silica and nutrients and sufficient silica-secreting organisms (largely sponges and radiolarians) to take advantage of the silica. The nutrient-rich, cold marine waters in turn require active oceanic circulation, probably as active as today. Modern-day ocean circulation is driven by cold water at the poles sinking and moving toward the equator. Shallower waters move pole-ward to replace these, and cold, nutrient-rich waters from the deep sea rise along the margins of the continents. Beauchamp proposes a similar pattern in the Late Permian, with sea ice near the Permian poles generating the cool waters that upwell along the northern margin of Pangaea.[17] Sea ice does not mean glaciers, and there is no evidence for polar glaciers after the mid-Permian. There are other indications that the poles may have been relatively cool during the Late Permian, even as equatorial latitudes continued to be quite warm. Following this line of reasoning, the collapse of the silica factories could represent the onset of the global warming proposed on land (chapter 6). Warming would melt the sea ice, ending the robust oceanic circulation and perhaps even degenerating into sluggish circulation. Silica and oxygen are less soluble in warmer waters, so rapid warming would reduce circulation and might inhibit the activity of silica-secreting organisms. Here the timing of the increase in sea level and the demise of silica are critical. If, as Beauchamp suggests, these events occurred in the Late Changhsingian, before the extinction, then we can eliminate the alternative possibility that chert deposition ended simply because the radiolarians and siliceous sponges that secreted the silica either became extinct or so reduced in abundance as to be ecologically insignificant.[18] Finally, the active circulation required by the chert deposition poses a considerable challenge to claims of pervasive low-oxygen conditions in shallow marine waters through the boundary interval. Such active circulation would seem to eliminate the possibility of global shallow-water anoxic layers. Deep-water anoxia is unlikely to be as severely affected.

The evidence from stable isotopes is abundant, consistent, and ambiguous. A large volume of organic carbon was dumped into the oceans and atmosphere during the end-Permian mass extinction. This shift has been found in many sections around the world, and as we have seen, shows up in both the organic and inorganic carbon reservoirs. The simple truth is that there are several equally plausible ways of interpreting the data from the carbon isotope record. If the carbon isotopic values of the entire ocean changed, which seems unlikely, then release of a large volume of methane is the most plausible alternative. New, smaller estimates of the amount of methane available today raise questions about the volume of sea-floor methane reservoirs available during the latest Permian. The realization that sea level was rising during the extinction causes a problem in understanding how the methane was released. Another source of methane is the heating of coals during the eruption of the Siberian volcanics. This took us to the alternative suggestion that at least part of the carbon excursion seen at the Permo-Triassic boundary was due to elimination of photosynthesis in surface waters, coupled with a reduction in the amount of organic carbon buried.

Two other isotopes also change during the boundary interval, sulfur and strontium. Although Kaiho has proposed a single explanation for this, via an extraterrestrial impact triggering massive flood-basalt volcanism, the change in the sulfur isotopes may reflect the growth of deep-water anoxia. To close the story of anoxia, we turned to Paul Wignall and Tony Hallam's work suggesting that shallow-water anoxia developed as sea levels rose in the Late Permian and Early Triassic, causing the extinction of shallow marine animals by slow strangulation. This datum conflicts with the silica factories found in the high Arctic, which can only form when vigorous oceanic circulation is bringing nutrient-rich waters up from the deep sea. It is now time to return to the various hypotheses I laid out in chapter 2, and to test them against the evidence we have developed in chapters 3–7.

CHAPTER 8
Denouement

I ended chapter 2 with three general questions that will help sort out the viable hypotheses from the moribund: Was the extinction rapid, even catastrophic, or a more drawn-out event? How well do environmental and climatic events correlate with the episodes of extinction? Finally, do the patterns of extinction match those expected from the various extinction scenarios, in particular are both the marine and terrestrial extinctions correlated? These questions will let us quickly dispose of several hypotheses and concentrate on the most persuasive.

I divided the various extinction proposals into six classes: (1) the effects of an extraterrestrial impact, (2) the climatic aftermath of the massive volcanic flood basalts in Siberia, and the possibility that an impact induced the eruptions, (3) extinction due to a decline in the number of biotic provinces, and species-area effects from the formation of Pangaea, (4) glaciation causing extinctions due to a combination of climatic cooling and a decline in sea level, (5) a variety of models involving the disappearance of oxygen from shallow and deep waters, including the Hallam and Wignall hypothesis of the movement of a shallow anoxic layer onshore with rising sea level, and the Knoll model of overturn of anoxic deepwater, and finally (6) the *Murder on the Orient Express* hypothesis suggesting that multiple interacting causes were responsible for the extinction. I have my own favorites among the various hypotheses, and some of my colleagues will doubtless disagree considerably with parts of this chapter. Moreover, if research continues at the rapid pace that we have seen over the past few years or so, new evidence may soon change these conclusions. To winnow these down to the leading hypotheses, let's begin with the three questions I posed at the end of chapter 2 and repeated at the beginning of this chapter.

The combination of the high-resolution radiometric dates from south China and the detailed statistical analysis of fossil occurrences has demonstrated to virtually everyone's satisfaction that the end-Permian mass extinction was very rapid in the oceans and probably occurred in less than a few hundred thousand years. Although debate continues about how much less than a few hundred thousand years, and there is still some concern over the precise age of the Permo-Triassic boundary, the rapidity of the extinction is not in doubt. Using the shift in carbon isotopes as a marker for correlation to other marine localities strongly suggests that the extinction was equally rapid in other parts of the world. In the absence of the detailed statistical work described in chapter 5 for Meishan, this conclusion is not as firm as one might like and the possibility exists that the pace of extinction could vary in different parts of the globe if we could resolve time finely enough. Finally, we believe that the identification of the carbon isotope shift in the terrestrial sediments of the Karoo and the interfingering of marine and terrestrial rocks in Greenland implies that the extinctions on land and sea were essentially coincident. This is an issue about which I continue to be concerned, and the discovery that these events were not precisely simultaneous would not greatly surprise me.

Since the hypothesis that the formation of the supercontinent of Pangaea was responsible for the extinction through a drop in the number of marine biotic provinces requires a fairly gradual extinction as the continents collide, such a rapid extinction effectively eliminates this scenario. We can in any case eliminate this hypothesis by noting that Pangaea actually formed in the Middle Permian, well before either the end-Guadalupian mass extinction, or our primary focus, the massive end-Changhsingian extinction that brought the Permian to a close.

The correlation between various changes in the physical environment and the pattern of extinction is the next general question. The Siberian volcanism is strongly correlated with the date of the marine mass extinction at Meishan and elsewhere in south China, so this hypothesis we will retain for more detailed consideration later in the chapter. The fourth class of extinction hypothesis invoked latest Permian glaciation, but this can be rejected on several counts. First, there is little evidence for massive glaciation at the Permo-Triassic boundary or on the timescale of the extinction. Second, the documentation provided by Tony Hallam and Paul Wignall establishes that in most regions, sea level was rising during the extinction interval rather than falling. Since the water for continental glaciers comes from the sea, sea level declines during a glaciation, and rises as glaciers melt. The absence of glaciation also removes the mechanism for oceanic overturn and carbon dioxide poisoning suggested by Andy Knoll and colleagues, but leaves untouched the suggestive pattern of extinction that might be explained by a different source of elevated atmospheric carbon dioxide levels. As I noted in the previous chapter, there is some evidence for cold poles in the very latest Permian, and Richard Bambach suggested to me that this could be enough to drive circulation and oceanic overturn.

Anoxia is a persistent feature of many different studies but has been damaged as a sole cause of extinctions on land among vertebrates, plants, and insects. These extinctions are difficult to explain by anoxia unless the volume of carbon dioxide released was sufficient to cause a significant drop in atmospheric oxygen levels. Advocates of these hypotheses have produced no quantitative models of the amount of carbon dioxide required to affect such an extinction, or plausibility studies that such an amount was available. Thus if anoxia was involved it must reflect a more general cause rather than the primary or exclusive cause. Finally, the extinction strongly correlates with a massive shift in the carbon cycle. While most geologists have viewed this as reflecting the cause of the extinction, we saw that at least a part of it may better be explained as a temporary consequence of the elimination of primary productivity during the extinction.

We have been able to reject both the biotic homogenization due to the formation of Pangaea and glaciation as participants in the extinction. We have three different models that involve anoxia in different ways. The Hallam and Wignall shallow water anoxia as well as Isozaki's deep-water anoxia models are expected to preferentially extinguish high-metabolic organisms. Yet the Knoll hypercapnia model suggests that just the opposite occurred, with groups with high metabolisms surviving better than those with low metabolic rates. The role of anoxia as a sole cause has been damaged by the extent of extinctions on land, which seem difficult to explain by anoxia. This leaves our search for the cause of the extinction focused on the effects of the Siberian flood basalt volcanism, the possibility of an extraterrestrial impact, with low oxygen in both shallow and deep marine waters as a contributory factor. The patterns of survival and extinction also suggest that we must pursue some variant of the hypercapnia model, albeit driven by something other than glacially induced oceanic overturn. Finally, we must also consider the very real possibility that the extent of the end-Permian mass extinction reflects several interacting processes rather than a single trigger. As fond as I am of this perspective, it too has been damaged by the rapidity of the extinction. Such a sudden event certainly suggests a single trigger.

· · ·

Volcanoes have many nasty and vile effects but are surprisingly unlikely to cause even a minor mass extinction. Our image of volcanic eruptions has been shaped by such recent events as Mount Pinatubo in the Philippines in 1991, seeing eruptions on Hawaii or Iceland on cable television, or stories of the burial of Pompeii by the eruption of Vesuvius in A.D. 79. The more than 5 billion tons of ash injected into the atmosphere by Mount Pinatubo cooled global temperatures for several months, and although it wiped out nearby plants, animals, farms, and one U.S. Air Force base, there was little far-reaching biological impact. In fact civilization has been fairly lucky over the past millennium with the absence of any really good-sized volcanic eruptions (luck that is sure to give out with a bang). Mount Pinatubo, Mount Saint Helens, and other recent eruptions were a fraction of the size of many older explosive volcanic events. But there is no good evidence that any of these larger volcanic eruptions have had any biological impact. Volcanic ash and gas can produce several years of cooler temperatures and better sunsets, but otherwise single, massive volcanic eruptions do not appear to have caused the extinction of anything, much less an extinction the size of the end-Permian.

But this applies to ordinary, run-of-the-mill volcanoes of the sort that ring the Pacific and pop up above zones where tectonic plates dive into the earth's mantle, or ooze out in the massive volcanism seen in Hawaii. Flood basalt eruptions are very different from the explosive, or pyroclastic eruptions of Pinatubo and Vesuvius. While they occur in the oceans as well as on land, continental flood basalts have received great attention because they seem more likely to cause sufficient climatic disruptions to trigger mass extinctions. And the really massive flood basalts such as Siberia and the Central Atlantic Magmatic Province (CAMP) that coincide with the end-Triassic mass extinction must have been produced so fast that even the smaller flood basalts are a poor analogue. Here we have no historical memory, no ready basis of comparison, only our imaginations.

No volcanic eruption on land in the past million years gives us an idea of what an "average" Siberian eruption may have been like. Let's revisit some of the statistics from chapter 2: Current estimates of the area of the Siberian flood basalts are about 7 million square kilometers (2.7 million square miles). Assuming the total eruption sequence lasted 1 million years, that means that on average 7 square kilometers (2.7 square miles) were covered each year, to a depth of sometimes as much as 3,000 to 6,000 meters (1.9 to 3.7 miles). And that is the problem. The eruptions did not slowly nibble away at Siberia but covered the same areas over and over. We know 45 separate volcanic flows have been identified from the 3,700-meter thick sequence in Noril'sk, for an average thickness of 82 meters (about 266 feet) for each flow (we have not accounted for volcanic tuffs and other types of volcanic rocks, but we will ignore them). A single eruption of Late Cretaceous Deccan flood basalt is estimated to have produced 1,000 cubic kilometers in a few weeks. At 82 meters thick, this would cover 8,200 square kilometers (about two-thirds of the State of Connecticut, or just less than the island of Puerto Rico), and 854 such flows would be required just to cover the area of the Siberian flood basalts. If we assume an average thickness of 2,000 meters, a total of almost 21,000 of these Deccan-sized eruptions are needed, or one every 48 years. On average, two-thirds of the state of Connecticut would be covered with 82 meters of basalt and volcanic debris every 48 years. In fact some eruptions covered vastly larger territories, so the frequency of eruption would have been less, but the ensuing destruction far greater.

At this point 1,000 cubic kilometers may seem small, so we need to put these numbers in the context of recent volcanic eruptions. The largest volcanic eruption of the past 10,000 years was the 1815 eruption of Tambour at about 100 cubic kilometers of volcanic material. This eruption cooled the earth by several degrees and triggered snowfall the following July in New Haven, Connecticut (best thing that could happen to Yale). The last volcanic eruption of near 1,000 cubic kilometers was the Yellowstone eruptions 1.3 million years ago. During the Siberian flood basalts, such eruptions would have been occurring twice a century.

There is the intrigin' issue, as Lord Peter Wimsey would have put it, of what a flood basalt actually does to kill off so many species. I remain amazed at how many scientists who should know better are perfectly accepting of temporal correlation, without inquiring into causal connections. Aside from the local destruction, the global effects of such eruptions must have been calamitous. Even if we assume that the cooling temperatures from the volcanic dust only lasted a decade after each eruption, add the effects of acid rain from the clouds of sulfuric acid and probably greenhouse warming from venting carbon dioxide, and only Dante could truly do this world justice.

Carbon dioxide is very effective at heating the atmosphere by retaining more of the Sun's heat on Earth rather than letting it radiate back into space. Volcanoes often emit large volumes of carbon dioxide as they erupt, and a recent paper by a German group suggests that carbon in the Earth's mantle may exist in a separate, carbonate pool. Extreme volcanic eruptions like the Siberian flood basalts could tap this pool and release huge volumes of carbon dioxide. Establishing how much carbon dioxide came from the Siberian flood basalts is impossible, and while such a source could contribute to a global greenhouse effect, remember that it could not produce the shift in the carbon cycle recorded in the carbon isotopic record. But recall that I pointed out that the Tugusskaya Series, through which much of the Siberian volcanics erupted, was the world's largest coal basin. How large it was before the volcanic eruptions vaporized a good bit of it is unclear, but as Gerry Czmanske has noted, this would have considerably enhanced the amount of carbon dioxide in the atmosphere. In chapter 7 I noted that heating the coal beyond 100°–200°C could convert coal to methane. Together these three sources suggest that the volume of carbon injected into the atmosphere both directly as carbon dioxide and as methane may have been very large, although firm estimates remain a subject for future research.

Extinction models have largely focused on the release of sulfate aerosols. Recall that sulfur commonly blows out of a volcano. In Siberia, the erupting flows may have destroyed evaporites rich in sulfates along with the coal beds. As the name suggests, evaporites are formed by evaporating seawater, leaving behind the various salts. Sulfur is an important part of some of these salts and vaporized evaporates would further increase the amount of sulfur in the atmosphere. Sulfates have the opposite climatic effect of carbon dioxide, cooling the atmosphere rather than warming it. Massive volumes of sulfates could cause a sufficiently long period of global cooling to form ice caps, triggering a drop in sea level. An abrupt end to the volcanism could cause melting of the ice cap and a rapid rise in sea level. There is not, however, evidence of either massive glacial formations or a drop in sea level. From this we can conclude that either the amount of sulfate released was not great, or that they were released over a longer period and never built up to great levels in the atmosphere.

A more important factor in the extinction may have been the conversion of sulfate aerosols to sulfuric acid and then acid rain. If volcanic eruptions are sufficiently explosive, the sulfates can be injected into the stratosphere where they can be carried far beyond the eruption and turn rain acidic. The devastating effects of much milder acid rain near power plants in the United States or the more severe destruction near industrial sites in the former Soviet Union might be a shadow of events at the Permo-Triassic boundary. Basaltic eruptions normally ooze rather than explode, so few of the gasses make it to the stratosphere, but 10%–20% of Siberian volcanic material came from very explosive volcanic eruptions. Prolonged release of sulfates may have translated into a lengthy period of acidic rain.

Finally, recall from chapter 6 that the evidence of mutations in lycopod spores from the Permo-Triassic boundary in India suggested to Henk Visscher that sulfuric acid from the acid rain reacted with chlorine to produce molecules that, along with bromine from the volcanic eruptions, destroyed stratospheric ozone. The mutations in the spores resulted from an increased mutation rate associated with the destruction of the protective ozone shield around the earth. The argument is speculative, as it requires interaction between the erupting volcanics and a hydrothermal system, and there are few constraints on how much chlorine and bromine may have been produced by the Siberian volcanics, or on the volume of such constituents required to cause a sufficient drop in stratospheric ozone.[1]

. . .

Establishing the origin of the Siberian flood basalt may help in determining whether the flood basalt was related to the mass extinction, but at the moment there is little agreement among geologists as to how flood basalts form in general, the Siberian basalts in particular. Since the early days of the revolution in plate tectonics, most geologists have followed J. Tuzo Wilson's argument that flood basalts represent a plume of material rising from deep in the earth. In 1963, Wilson, one of the founders of plate tectonics, suggested that the movement of the Pacific tectonic plate formed the Hawaiian Islands as it drifted across a persistent conduit of magma rising from deep in the mantle. Islands grew from these subsea eruptions, and as the plate moves, one center of eruption dies off and another begins. Because the movement of the plates did not appear to influence the plume, the source of it was thought to be fairly deep within the earth, with the plate sliding over the top of a plume much as your hand might over a candle. Other geologists later suggested that the plumes were contained within fairly narrow conduits ascending from close to the boundary between the earth's core and mantle. In 1981 Jason Morgan suggested that the plume responsible for the Siberian flood basalt left a track northward through the Arctic Ocean along the Lomonosov Ridge and today lies beneath Iceland. But when the ridge was sampled a few years later geologists found not volcanic rocks but a rifted piece of the European continent. More recently other geologists have identified another possible plume track through a trough south of the Taimyr Peninsula, but this purported track seems unlikely as well.

Despite the near-universal claims of geology textbooks, no one has been able to demonstrate that plumes arising from deep within the earth actually exist. This view of plumes as a chimney of material rising from the deep mantle may be utterly fallacious. By plotting thousands of earthquakes and their paths through the earth, seismologists can produce a sort of CAT scan of the earth (which goes by the name mantle tomography). The waves produced in an earthquake slow down as they hit hotter regions inside the earth and this variation in the speed at which the waves travel allows seismologists to produce images of mantle plumes. In Iceland, for example, the plume only extends down 400 kilometers, rather than the 2,900 that would be expected if the plume began at the base of the mantle. Similar studies are under way in Hawaii and Yellowstone, but enough work has been done to raise questions about whether deep sources of these supposed mantle plumes exist. Of the forty-five to fifty generally accepted hot spots, only seven or eight have plumes known to go deeper than 200 kilometers.[2]

Whether a mantle plume was associated with the Siberian flood basalts is a contentious subject within certain geological circles. The flood basalt erupted through a Middle Carboniferous to Late Permian suite of river-borne conglomerates, limestones, and coals known as the Tugusskaya Series. As mentioned in chapter 7, the coals comprise the world's largest coal basin. The Russian geologist Valeri Fedorenko, Gerry Czemanske of the U.S. Geological Survey, and their colleagues have conducted incredibly thorough

studies of the region, mapping these rocks, measuring their thick-
nesses, and determining the environments in which they formed.
The maps they produced from this data show how Siberia evolved
during the 100 million years before the end-Permian mass extinction: A succession of forests and peat swamps formed the coal, but
after the coal was deposited, the region was uplifted, eroded, and
then returned to sea level. Marine fossils are found between the
flows in the first kilometer of the flood basalt volcanics, showing
that they erupted near sea level. In fact, their detailed geologic
mapping demonstrates that Siberia was slowly subsiding during
most of the Permian, and continued to subside during the volcanic eruptions.[3]

This evidence for slow subsidence during the eruption is the
opposite of expectations from a mantle plume. The immense volume of very hot material in a rising plume should pool beneath
the crust before an eruption, raising the continent above. Such
continental uplift can be a good clue to the presence of a mantle
plume. Two hundred fifty million years later, the uplift will long
since have disappeared, but it leaves a sort of geological fingerprint in the faulting around the margins of the uplifted area and
in the deposition of sediments eroded off the uplift and carried
away in rivers and streams. The extent of the uplift will depend
on the size of the plume head, but a plume with a 400-kilometer
radius should cause uplift of 2–4 kilometers for perhaps 5–20 million years. As the volcanic eruptions begin, the uplift will subside.
But the geologic mapping of Czemanske shows that Siberia began
subsiding before the eruption. Evidently the accumulation of the
volcanic material was balanced by subsidence, so the overall topog-
raphy was essentially flat.

Other geologists, examining different data, have arrived at a
different conclusion from that of Czemanske and colleagues. A
group at the University of Leicester in England headed by Andrew
Saunders and Marc Reichow has analyzed seismic sections, rocks,
and other data. From the chemistry of the rocks they concluded
that the basalts are fairly primitive, meaning that they are very
similar to the Earth's mantle. They also concluded that the West
Siberian region was uplifted prior to the eruption of the flood

basalt. This is hard to reconcile with the far larger and more detailed studies of Czemanske, Fedorenko, and their colleagues.
While it is possible that the locus of the mantle plume was to the
west of the classic Siberian flood basalt region, the apparent absence of uplift under the region with the bulk of the volcanics
seems a bit odd, at least to me.[4]

I have been describing the fairly classic plume head model:
enormous 500–2,000-kilometer-wide blobs of mantle material sitting atop a long tail extending deep into the mantle. An interesting alternative is a plume with no blob on top, the plume tail
model. The tails are perhaps 100 kilometers in diameter and consequently would cause little uplift of the overlying crust. Lindy
Elkins-Tanton, a recent graduate student with Brad Hager at MIT,
and now at Brown University, showed that this fits the data for
Siberia quite well.[5] The uplift and erosion between the coal beds
and the volcanics could mark the first appearance of the mantle
tail. The mantle tail would heat and "erode" the base of the crust
as the volume of melt increased. Elkins-Tanton and Hager showed
that this mechanism could produce the required volume of material in less than 1 million years and would produce the gradual
subsidence observed by Fedorenko and colleagues in the field.

Mantle plumes might also be a red herring, intellectual detritus
from the early days of plate tectonics. The chemistry of the basalts
and other rocks in Siberia suggests an incipient attempt to tear
Asia apart along a continental rift system, much as the separation
of Africa and South America during the Mesozoic created the Atlantic Ocean. The chemistry of most continental flood basalts is
similar to oceanic islands like Hawaii. Rocks from island arc volcanoes above a subducting plate, such as those in Japan or Chile,
have a very different chemical signature. The Siberian flood basalts, the end-Triassic Central Atlantic Magmatic Province, and the
Lesotho flood basalts in South Africa are more similar to island arc
basalts than other flood basalts, however, suggesting that continental rifting is a more plausible explanation than a mantle plume
for these eruptions. Both this rifting hypothesis and Lindy Elkins-
Taunton's plume-tail model share a potentially significant difficulty: they require sufficient heat flow to melt the lower crust and

to produce the huge volumes of melt. As I mentioned, Lindy Elkins-Taunton's computer simulations suggest this is possible, but Marc Reichow's group reached just the opposite conclusion. While I was discussing these issues with Lindy in early 2005, I suddenly realized that all large continental flood basalt provinces are either on the margins of tectonic plates or along rifting zones. This makes sense from the need to have high enough heat flow to produce the large volume of melted rock that produces the erupted basalt. The sole exception to this generality is Siberia: an enormous flood basalt in the midst of a massive continent. Or is it an exception?

Vincent Courtillot and his colleagues at the Institut de Physique du Globe de Paris found that the eight major continental flood basalts of the past 300 million years are associated with continental rifting.[6] For example, flood basalts in Ethiopia 30 million years ago accompanied rifting in the Red Sea and the Gulf of Aden while flood basalts in the Parana of Brazil correspond with the opening of the South Atlantic. Yet both the Ethiopian and Parana flood basalts have chemistry similar to other flood basalts, and not to Siberia, the Central Atlantic flood basalts, or the Karoo. For the Siberian flood basalts the association with continental rifting is less obvious. Courtillot and colleagues suggest that the Tugusskaya basin reflects a failed attempt to rift the continent. But this conflicts with Fedorenko and Czemanske's careful geologic mapping showing that the oldest rocks in the Tugusskaya date to the middle Carboniferous, making the basin much older than required by Courtillot's model.

Time has not dealt kindly with other elements of the proposed association between flood basalts and the extinction. Better radiometric dating of both extinctions and volcanic episodes now reduces the number of paired events from nine to four and this includes the Emeishan volcanics from China. As I noted in chapter 4, since we do not know the date of the end-Guadalupian extinction event, the only basis for correlating the extinction and the volcanics is that the volcanics sit on top of the Guadalupian-age Maokou Formation. So they roughly coincide, but better stratigraphic correlation or coincident radiometric ages would be useful. More troubling are observations that the eruptions generally begin after the mass extinction. There is also a lack of any obvious relationship between the extent of the mass extinction and either the magnitude of the volcanism or the apparent degree of sulfates released.[7] Any association between volcanism and mass extinction must be indirect, perhaps with massive volcanism triggering runaway greenhouse effects and release of methane hydrates.

So what did cause the Siberian flood basalts? There is little evidence to support a significant mantle plume. While the plume-tail hypothesis is attractive, the rocks are chemically different from most other large flood basalts. The fact that other flood basalts are associated with either rifts or the margin of mantle plates, along with some of the chemistry of the Siberian volcanics, suggests a rifting continent rather than a flood basalt, but this remains speculative. A very different explanation is that Siberian volcanism began not in the earth, but in the sky, with the impact of a meteor, to which we turn next.

. . .

My favorite claim of a connection between volcanism and impacts is the suggestion that the precision impact of several extraterrestrial objects in the midst of Pangaea triggered a great rift, which gradually widened into the opening of the Atlantic during the Late Triassic and Jurassic. Remarkably, this resurfaced at the 2002 spring meeting of the American Geophysical Union with a proposed impact structure encompassing all of northern Africa and centered off Cameroon[8]—a theory enveloping the end-Permian mass extinction and the fragmentation of Pangaea, and offering a new driving force for plate tectonics. No doubt Hollywood has optioned the movie rights.

Opponents of the impact hypothesis seized upon the apparent eruption of the Deccan flood basalt in India at the Cretaceous-Tertiary boundary to argue that it was the cause of the extinction. Many geologists have compared the dates of the past 300 million years of major continental flood basalts to the ages of mass extinctions (figure 8.1). The end-Permian, end-Triassic, and end-Cretaceous mass extinctions all coincide with significant flood basalts, the first two with the two largest continental flood basalts in the

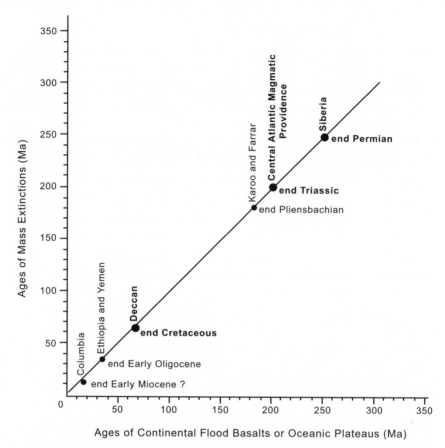

Figure 8.1 The correlation between radiometric ages of major flood basalts and several mass extinctions is remarkably good.

past 600 million years. Several smaller biotic crises of the past 250 million years also correspond to flood basalts ages. If that is a coincidence, it is certainly a curious one. It may be that closer inspection will reveal that the dating of either the flood basalts or the extinctions are off, but the correlation between the end-Permian mass extinction and the Siberian flood basalts are growing stronger. This correlation between extinctions and flood basalts has led geologists to investigate whether they could have been triggered by impacts.

The critical questions are how large an impact is required to cause volcanism, how much volcanism can be expected, and, if impacts can induce volcanism, whether there is any good evidence that impact caused the Siberian flood basalt. And by good evidence I mean more than "we can expect X to occur once every 250 million years; the end-Permian mass extinction occurred 250 million years ago; ergo X caused the extinction."

Addressing the question of whether impacts can induce flood basalts seems fairly straightforward, potentially even fun: wander around the world, check which flood basalts sit in impact craters, voilà. Regrettably rain, wind, and movement of the plates have obliterated many impact structures. After all, the Chixulub impact crater is only 65 million years old and there is only a faint trace of it at the surface of the Yucatan peninsula. The late Gene Shoemaker of the U.S. Geological Survey, and one of the founders of planetary geology studied asteroids and comets whose orbits cross that of the earth and also counted the cratering rate on the moon. He concluded that we should expect an impact producing a 10-kilometer crater every 4 million years or so (give or take 2 million years), or about 516 such craters in the past 120 million years. Correcting for the greater surface area of the oceans, 212 of these impacts should have hit the continents, but only 89 craters over 10 kilometers in size have been discovered. Despite weathering and other uncertainties we still seem to be missing quite a few craters. Incidentally, it was this research that changed many geologists' views of the end-Cretaceous mass extinction, including mine. Many geologists opposed the Alvarez hypothesis in 1980 because of its whiff of catastrophism. Shoemaker removed this by showing that over a longer time scale the impact of extraterrestrial objects was expected.[9]

Determining the date of impact craters is a more indirect approach to searching for a connection between impacts and flood basalts. These dates can then be compared to the dates of flood basalts, as in the evaluation of the correlation between flood basalts and mass extinctions. A recent comparison of both lunar and earth impacts provided a cratering record of the past 3.8 billion years, assuming that the moon's craters could stand in for the missing craters on Earth. The comparison with mantle plumes at first seems impressive, but the dates are sufficiently coarse that they must be smoothed by binning them into 30- to 45-million-year intervals. The suggestion that an impact and flood basalt occurred in the same 30- to 45-million-year interval is hardly compelling. An even more serious flaw is the inclusion of all impacts, not just those large enough to trigger volcanism. The computer simulations I will discuss shortly suggest that only objects capable of producing craters larger than 250 kilometers (and perhaps as large as 500 kilometers) should have been included. Only two to four such continental craters are likely to have occurred in the last billion years. With thirty flood basalts in the past 250 million years, we would seem to have too many flood basalts, or too few large-impact craters.[10]

Unlike Earth, weathering on the moon and Mars is insignificant. A second approach is to look around the solar system for volcanism triggered by impact. Well before the Apollo moon program, astronomers studying the moon suspected that the dark material filling many impact basins was basalt. The first satellite images of the far side of the moon showed basins that were not filled with basalt, and the Apollo missions confirmed that while some basins are filled with basalt, the basalts are as much as a billion years younger than the basins. Jay Melosh at the University of Arizona is a leading expert in this area, and after surveying craters on the earth, the moon, and Venus, he concluded that there is absolutely no evidence that impacts have caused volcanism anywhere else in the solar system. Impacts produce vast sheets of melted rocks but these melts are nothing like lavas; they are not found with volcanoes and do not approach the volume of large flood basalts on earth.

The nuclear arms race between the Soviet Union and the United States employed vast armies of physicists in building computer models of impacts and explosions, models that can be converted for the more socially useful question of what happens when a meteor hits the earth. Working with B. A. Ivanov of the Institute for the Dynamics of the Geosphere in Moscow, Melosh developed a computer model of a 20-kilometer-diameter meteorite striking the earth. The resulting 250–300-kilometer-diameter crater produces about 10,000 cubic kilometers of melted rocks, and a flat crater. Nary a smidgen of volcanism, sad to say. Depending on the temperature of the target rock (hot rocks are easier to melt than cold ones), Ivanov and Melosh find that a crater diameter of 500–1,200 kilometers is required to trigger volcanic activity. No impact structures this large have been found on the Earth or Venus, and the few on the moon are older than 3 billion years, representing the final cleaning of debris during the formation of the solar system.[11]

A nifty, related idea is known in the trade as antipodal focusing: the energy of an impact is focused on the opposite side of the earth from the point of impact. Both large impacts and large earthquakes cause the earth to ring like a bell. According to antipodal focusing, when the energy waves reach the opposite side of the earth the collision of the waves will focus energy and could trigger volcanic eruptions. But with fairly simply physics Jay Melosh showed that the amount of energy available through this process would not cook a hot dog: there simply is not sufficient energy available to trigger antipodal volcanism.

Lindy Elkins-Tanton and her MIT advisors Brad Hager and Tim Grove used computer models to investigate the effects of an impact but came to slightly different conclusions. Unlike Ivanov and Melosh, their models suggest that large lunar impacts could trigger volcanism on the moon, and that the volcanism could continue for as long as 350 million years after an impact. This would account for some of the discrepancies between the dates of the basaltic "seas" on the moon and the impact craters in which they are found. Large volumes of melt rock could be produced by impact (indeed other recent computer simulations have reached the same conclusion). The unique contribution of Elkins-Tanton's work is the suggestion that as an impact crater forms, the underlying material will rise to replace what is blasted out of the crater, causing convection within the mantle, and volcanism. Differences between the moon and Earth also allow the volcanism to continue much longer on the former than it would on the latter.[12]

The same process may work on Earth, at least for impacts near the margins of the continents. As an illustration, a meteorite 15–25 kilometers in diameter produces a roughly 300-kilometer-diameter crater (depending on the speed of the impact and the mass of the object). The resulting crater would have a maximum depth of about 22.5 kilometers, and 1.2 million cubic kilometers of rock would be blown out of the crater, of which 100,000 cubic kilometers would provide the melt rocks. But this melt is less than 10% of the volume of the Siberian flood basalt and in any case is not basalt. Forming the crater reduces the pressure on the underlying rocks, however, and under the right circumstances this pressure release can cause additional melting deep below the crater. If the underlying mantle is hot enough, the MIT group showed that a 300-kilometer-diameter crater could produce 2–6 million cubic kilometers of basalt, just the volume needed for the Siberian flood basalts. Critics will doubtless respond that these are just computer simulations. True, but such simulations have become very sophisticated since the end-Cretaceous impact was proposed.

I have not raised the possibility that iridium, shocked quartz, spherules, or other impact debris came from volcanism, because the consensus developed among impact researchers is that pressures produced by volcanic eruptions are too low to produce shocked quartz, and the chemistry of the Earth and volcanoes was wrong for other impact signals. Since I find consensus as mind numbing as I do coincidence suspicious, I can happily report that J. Phipps Morgan recently assaulted this fortress of consensus.[13] Struck by the correspondence between flood basalts and mass extinctions, Morgan and his colleagues searched for a terrestrial process that could produce shocked quartz, microspherules, and other impact evidence. They proposed that large carbon-rich pools build up beneath continents and then erupt explosively, producing pressures more similar to impacts. These eruptions vent large volumes of carbon dioxide, sulfates, and debris into the stratosphere. This amendment offered to the standard Siberian story has the advantage of producing some impact signals. Unfortunately shocked quartz is, as we have seen, all too rare at the Permo-Triassic boundary. There is no evidence that fullerenes and microspherules could be formed in this way. Kimberlite pipes produced by supersonic jets of material from the mantle, and the source of diamonds, are found in the eastern region of the Siberian flood basalts, and this is consistent with Morgan's hypothesis. This new hypothesis is, as the authors admit, of the school of extreme geology. It will be interesting to see how the geologic community receives it.

Correlations between impacts and flood basalt, studies of impact craters on the moon and some computer simulations all raise doubts about impact-induced volcanism, at least on the scale of the Siberian flood basalts. But they have not completely excluded the possibility. An impact large enough to trigger volcanism requires a crater at least 250–300 kilometers in diameter. It seems impossible that such a massive crater would not have produced far more signs of impact than the Chixulub event, including massive amounts of shocked quartz. Yet the signs of impact at the Permo-Triassic boundary are far more circumstantial: the disputed fullerenes and possible meteorite fragments. Moreover, the Siberian flood basalts erupted from at least four different centers over 1,000 kilometers apart, a fact inconsistent with impact. We can, I think, safely conclude that impact is an implausible cause of the eruption, and that Tanton and Hager's mantle-tail hypothesis (the stealth-mantle plume) is the most likely cause of the flood basalt.

・　・　・

Just because the Siberian flood basalts were not caused by the impact of an extraterrestrial object does not mean that no impact occurred. Although I remain constitutionally opposed to coincidence (it is a lousy way to bet), it is possible that an impact occurred as the Siberian flood basalt erupted.

The rapidity of the marine extinction is consistent with an impact, but that hardly counts as evidence of impact, since other causes could produce a similar pattern. Tests for iridium, shocked quartz, and other generally accepted indicators of impact have come up negative. As discussed in chapter 2, Luann Becker's fullerenes are claimed to have trapped helium that must have come from outside the earth. Since the first paper reporting fullerenes from Meishan, South China, was published in 2001, Bob Poreda and Luann Becker have reported similar evidence from a Permo-Triassic boundary section at Graphite Peak, Antarctica, and from a more problematic section in Japan.[14] They found fullerenes in the "upper" boundary breccias at Graphite Peak with enrichment of helium-3 gasses. Samples from about 145 centimeters below the boundary lacked any trace of fullerenes or enriched helium gases. Helium-3 can come from other extraterrestrial particles, in particular from something known as interplanetary dust particles, or IDPs. IDPs are extremely small grains that float down continuously upon Earth and contain a distinctive chemical signal of the early solar nebula. After comparing the Graphite Peak data to a recent sample of known IDPs, Poreda and Becker found that they were very similar, suggesting that the enrichment of helium-3 may reflect an IDP origin and not an extraterrestrial impact.

Sam Bowring and I collected the samples studied by Becker from the south China sections. Sam gave samples from the same bed to Ken Farley at Caltech, who could find no sign of helium in the rocks. The beds from Meishan are complex and vary in composition between the different quarries and even within the same quarry. We have no certainty that Becker and Farley analyzed identical samples; when combined with the considerable difficulties in detecting fullerenes, the failure to replicate Becker's results may not be surprising. Becker has claimed that fullerenes are highly resistant to destruction by geological processes, but they are difficult to isolate from rocks where they should occur and are involved in many chemical reactions (the reason chemists are so entranced by them) so they may be less robust than claimed. Until Becker and Farley analyze exactly the same samples, or another lab replicates Becker's results, interpreting the reports of extraterrestrial fullerenes will remain difficult.

Far more suggestive is the possible Bedout impact structure off western Australia, also proposed by Luann Becker. Here we have a structure that on admittedly coarse imaging and gravity data has been interpreted as an impact structure, putative impact debris recovered from cores of ocean-bottom sediments, and a single radiometric date close to the Permo-Triassic boundary. I am not a specialist on impact evidence, but in talking to several such experts since the Becker paper was published, it is clear that they are less than impressed by this evidence. The nature of the structure is too poorly defined to provide much real basis for support or rejection of an impact origin, and so much of the criticism focuses on the rocks recovered from the structure. The shock of an impact can profoundly alter the mineralogy of the surrounding rocks. Becker and her colleagues found many examples that they interpreted as signs of impact. Such members of the impact community as Bevan French in my department at the Smithsonian, Cristian Korberl of the University of Vienna, Richard Grieve of the Canadian Geological Survey, and Andrew Glikson of the Australian National University have all discounted this evidence. In October 2004, *Science* published three responses critical of the Bedout structure. Glikson has examined the same drill core, for example, and interprets the rocks as a volcanic breccia with no evidence of shock or impact. He was particularly suspicious of the lack of minerals with measured planar-deformation features, a fracture pattern characteristic of impacts. Paul Wignall and colleagues have been studying the Hovea-3 drill core taken from about 1,000 kilometers distant from Bedout, where there is no evidence of impact debris, elevated trace elements such as iridium, or disruption at the Permo-Triassic boundary, all of which casts considerable doubt on Bedout as a Permo-Triassic impact. In their response Becker and her colleagues raised some criticisms of their own, pointing to the unusual chemistry of some of the rocks, the issue of how volcanism would have arisen along the tectonically inactive western margin of Australia, and they reiterate their claim for an impact source for the Bedout crater. Finally, the single radiometric date at the boundary is problematic. In their comment, Paul Renne and others raised questions about the date and conclude that the data presented in the original paper do not support the claimed date of 250.6 +/− 4.3 million years. Although Becker's group disputes this, they admit that further radiometric dating is required. The Bedout structure is certainly sufficiently interesting to warrant continued investigation, but at present it provides little real support for an impact at the Permo-Triassic boundary.[15]

Support for an impact grew stronger in October 2003 with the identification of forty potential meteorite fragments and metal grains from the Permo-Triassic boundary at Graphite Peak in Antarctica.[16] The Graphite Peak section is where Retallack reported some questionable shocked quartz, and Becker's group identified fullerenes. The grains from Graphite Peak include iron-rich nuggets, iron-nickel-phosphorus-sulfur minerals, and rare iron-sulfur

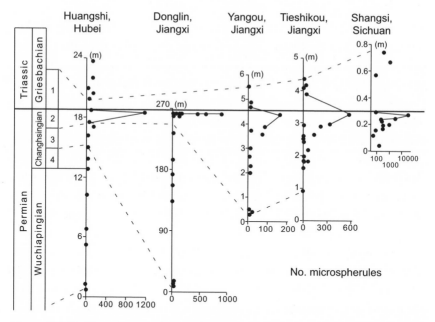

Figure 8.2 Jin Yugan's group in Nanjing has looked for microspherules at a many different Permo-Triassic boundary localities in China. In this figure, the dots show the number of microspherules in selected beds near above and below the boundary. Note that the microspherules are abundant only in late Changhsingian beds near the boundary. Localities show the city and province nearest the localities, dotted lines correlations between sections.

minerals. This suite of minerals is unusual on Earth, but matches a group of meteorites known as Mighei-type carbonaceous chondrites, named for a meteorite that fell in the Ukraine in 1889. Carbonaceous chondrites are a rare form of stony meteorite, less than 3% of all meteorite finds, but scientists studying meteorites as primitive samples of the nebula from which the solar system formed particularly prize them. Mighei-type chondrites are rich in organics and minerals formed in the presence of water. The same research group also recovered similar iron-nickel grains from Meishan, in the rusty layer at the contact between beds 24 and 25, similar to grains reported since the 1980s (figure 8.2). These are almost pure iron with very small (less than 1%) amounts of nickel, chromium, and silicon. Such striking similarity from Antarctica (which was near the south pole in the Permian) to the equatorial latitudes of China requires some means of global distribution, and similar particles could yet be found in many other boundary rocks.

How such small and incredibly fresh-looking metallic fragments were preserved for 251 million years is a real puzzle. The entire geological literature contains only three other reports of similarly minute meteorite fragments (some 480-million-year-old meteorite fragments are several inches across). Exposure to water and air normally turns such grains into clay particles. It is possible that quick burial in the anoxic environment of the boundary clays helped to preserve them. Contamination is one of the first concerns that scientists have in such situations, but the researchers took extensive precautions to guard against this by burrowing beneath the surface at least six inches and analyzing the samples in a lab clean of previous exposure to meteoritic material.

Evidence for an impact at the Permo-Triassic boundary has been growing over the past few years. The demonstration that the rate of extinction was very sudden and perhaps catastrophic, that the marine and terrestrial events were very close in time, that the "fungal spike," which began about 500,000 years before the extinction, was not actually fungal but algal (and thus can be discounted as evidence of early terrestrial disturbance), the shift in the carbon cycle, the discovery of fullerenes with a solar nebular signal, and most recently the micro-meteorite fragments from Antarctica and Meishan are all suggestive of an extraterrestrial impact.

Yet doubts persist, and for good reasons. Why has strong evidence of impact proved so elusive? The end-Permian mass extinction was a far larger than the end-Cretaceous mass extinction and what triggered it must have been significantly more powerful as well. The success of the Alvarez hypothesis may have blinded us to the diversity of impacting objects so that we have been searching for evidence of an object like the one that hit the Yucatan Peninsula 65 million years ago, rich in iridium and producing shocked quartz. Other impacts might leave very different signatures. There is no reason a comet, for example, would contain much iridium. In 1998 Sam Bowring and I suggested that the abundance of very negative methane in some comets could explain the sharp negative carbon shift at the Permo-Triassic boundary.[17] It is unclear whether an impact could have triggered the Siberian flood basalts, but if it did it must have been massive. The lack of impact debris or shocked quartz, the geographic extent of the volcanic source areas, and the absence of gravity or magnetic evidence of such a large impact crater all argue against an impact origin for the Siberian flood basalts.

I have a final and utterly nonscientific reason for hoping that impact was not the cause of the extinction. It would be boring. Many geologists and paleontologists have been hoping for a "general theory of mass extinctions," with all events linked to a common cause, whether glaciation, impact, flood basalts, or the spread of anoxic waters. I can see the attraction of such an argument, but it seems rather limiting, and particularly so in the case of the end-Cretaceous and end-Permian events. A decade ago these two events seemed utterly dissimilar: a rapid, catastrophic event versus a much slower episode. Today we see increasingly similar patterns, and some take this as a heartening sign that similar causes underlie both events. Contrarian that I am, I think we would learn more about how the earth operates if the extinctions had different causes. A very similar outcome from distinctly different triggers is far more informative about how the earth and life respond to environmental insults. If ecosystems on land and sea collapsed in similar ways, if similar changes occurred in the cycling of carbon and other elements and in climatic change, but were induced by very different processes this would open a new window for exploring how the diversity of life on Earth is regulated, and how life has persevered for so long.

· · ·

As we saw in chapter 7 both sulfur isotopes and the appearance of black shales and thinly laminated sediments in the shallow ocean and black cherts in deep water suggest that anoxic waters spread in both the deep ocean and in shallow marine environments near the Permo-Triassic boundary. This evidence has led to a plethora of different hypotheses linking such anoxic conditions to the mass extinction.

Paul Wignall and Tony Hallam proposed that a layer of low-oxygen waters moved onshore during the latest Changhsingian

transgression.[18] The disappearance of Permian reefs during the extinction could reflect the development of shallow water anoxia,
since latest Permian reefs are dominated by sponges and corals, a
pattern found today only on reefs in low-oxygen environments.[19]
Critical to their model is a tight association between the extinction
horizon and the appearance of other evidence of anoxia, such as
black shales. Paul and Tony have presented evidence that the actual
timing of extinction can differ by a few conodont zones from place
to place, tracking the sweep of these low-oxygen waters onshore.

Black shales are not the only indicator of shallow-water anoxia
used by Wignall and Hallam, but they are a significant one. Sedimentologists have long used black shales as a signal of the deposition and preservation of organic-rich carbon in anoxic settings.
But what geologists all know to be true may not be. Studies of Late
Cretaceous black shales from the western United States raise the
possibility that black shales may be formed by the adsorption of
carbon onto the surfaces of clay minerals. The black shales reflect
the production of clays through continental weathering rather
than oxygen in the waters or the magnitude of primary productivity.[20] This process of course would not apply to carbonate boundary sections, but some black shales at the Permo-Triassic boundary
could simply indicate increased clay deposition from volcanism or
greater continental weathering. Greater continental weathering
would be consistent with the acid rain from Siberian volcanism.

What of the other evidence of shallow-water anoxia? I have always been somewhat dubious about the thinly laminated earliest
Triassic sediments as evidence of anoxia. The laminated sediments
could simply mean that most burrowing organisms went extinct
or were very rare, and so few sediments were burrowed. Finally,
the Wignall and Hallam anoxia model provides no explanation
for the disappearance of plants, insects, and vertebrates on land.
I hasten to add that I do not want to cast too much doubt on
the existence of shallow-water anoxia during the earliest Triassic,
because I do believe the evidence is reasonably suggestive. I am
less sure that this was sufficient to explain the magnitude and pattern of the mass extinction.

Andy Knoll, and his colleagues John Grotzinger, Richard Bambach, and Don Canfield, synthesized a broad array of information

in proposing that a rapid overturn of deep anoxic waters released a large volume of carbon dioxide into the atmosphere and triggered the extinction.[21] The most suggestive evidence for this release are those odd carbonates found near the boundary, so similar to rocks more than 600 million years ago and so unlike more modern rocks. This model explains both marine and terrestrial extinctions through carbon dioxide poisoning, or hypercapnia. As I have already noted, the invocation of glaciation to drive the overturning of the oceans is almost certainly incorrect, but it is possible that the cooling at the poles documented in the Sverdrup Basin might be sufficient to cause cooler waters to sink and trigger oceanic overturn. This modification of the Knoll hypothesis still faces the problem of whether enough carbon dioxide could have been sequestered in the deep sea to cause hypercapnia. All this does not negate the possibility that carbon dioxide was released by some other means. The buildup of carbon dioxide in the deep sea is consistent with Isozki's finding of anoxic sediments beginning early in the Late Permian, but of course these anoxic cherts persist well into the Early Triassic, long after the Knoll model suggests oceans should have become better oxygenated following overturn of the oceans. The sulfur isotopes provide a possible solution, as the Knoll group realized. The short reversal in the anoxic signal in the sulfur isotopes suggests a very brief oxygenation of the deep ocean just at the Permo-Triassic boundary. (A convulsive oceanic-overturn by an impact? One wonders.)

Despite what one might hear from certain politicians, computer climate models have become extraordinarily sophisticated in the past decade and are increasingly useful for really interesting questions like the climate of the Late Permian. Two different groups of climate modelers, one at University of Pennsylvania and a second at MIT, modeled Permian oceans to evaluate the Knoll oceanic-overturn hypothesis. The modelers were specifically concerned with testing whether a long-lasting, stagnant deep ocean could build up a sufficiently large volume of hydrogen sulfide and carbon dioxide to cause a mass extinction.[22] The two teams reached very different conclusions.

The abundant plants and animals in shallow oceans derive their energy from sunlight but require a constant flow of nutrients.

Some nutrients enter from rivers or estuaries, but the upwelling of cold, nutrient-rich waters from the deep sea is also vital. Such upwelling off the coast of Peru ends with the El Niño, named for the collapse of the anchovy fisheries near Christmastime. Photosynthetic plankton are abundant near upwelling zones, making them a good area to look for oil, so the petroleum industry has a strong interest in locating areas of past upwelling. The Penn State group found that a stratified ocean sharply reduced upwelling of nutrients from the deep sea. The resulting drop in shallow-water marine productivity could correspond to sequestering a large volume of carbon dioxide in the deep ocean during the Late Permian. In their computer simulations, the Penn State group imposed a lower temperature gradient between the equator and the poles than exists today. The warmer poles and slightly cooler equator reflect evidence from oxygen isotopes, fossil soils, and patterns of sediments. The results show that with a low temperature gradient between the poles and the equator, and limits to levels of photosynthesis, oceanic circulation is very sluggish, and the amount of oxygen in the deep ocean drops by more than three-quarters. Despite this remarkable drop in oxygen levels the reduction is not sufficient to produce the pattern required by the Knoll model.

The MIT model focused more on anoxia than upwelling and considered two different styles of oceanic circulation: The first model involved circulation similar to today, with cool waters near the South Pole sinking and bringing oxygen into deep waters; under such conditions they found that anoxia cannot develop in the deep sea. Their second model reverses the circulation with evaporation near the equator causing denser, salty water to sink into the deep ocean. In this case ocean stratification is pervasive and anoxia can develop. Yet the resulting anoxia will not persist for millions of years and it is unclear whether the degree of anoxia required by the Knoll mode could develop in a Late Permian setting.

The Penn State model is driven by changes in the pole to equator temperature gradient with oceanic circulation similar to today while the MIT model contains two distinct styles of oceanic circulation. While there are other technical differences responsible for some of the variation in results, both models can produce deep oceans with a significantly lower amount of oxygen but neither simulation could produce the persistent, deep-water anoxia that Isozaki has reported from geological evidence, that the sulfur isotope suggests occurs from the early Late Permian into the Early Triassic, or that the Knoll & Co. model requires. The resolution of this conundrum may lie, at least in part, in recognizing that we have been making a major assumption that may not be justified: that deep-sea sediments were a faithful record of the overlying waters. What if the sediments had little oxygen, but not the deep ocean waters? Decay of organic material settling through the ocean removes oxygen and often leaves little behind for oxygenation of the underlying sediment, particularly if there is little burrowing. This would produce the geological evidence but neither the deep-sea anoxia of Isozaki nor the vast buildup of carbon dioxide required by the ocean overturn hypothesis. It would also be consistent with the vigorous circulation required by the massive chert factories Benoit Beauchamp found in the Sverdrup Basin of Arctic Canada.

Previous chapters have mentioned changing style of rivers in the earliest Triassic, the end of the Permian Chert Event, and other evidence for climatic warming. This evidence is not yet overwhelming but new evidence continues to develop for a widespread greenhouse episode. Earliest Triassic fossil soils in Antarctica have an unusual green mineral known as berthierine. Both lab experiments and field evidence confirm that this forms in low oxygen settings. This could have formed as soil oxygen was consumed by massive amounts of methane, converting the methane to carbon dioxide and producing the greenhouse.[23] Again, I am far from clear as to whether this global warming reflects the cause of the extinction or its immediate aftermath, and possibly the cause of the delayed recovery in the Early Triassic.

The evidence for an earliest Triassic greenhouse has generated considerable enthusiasm for the release of large volumes of methane gas. I would take some of the credit (or blame) for this if it were not for the rather amusing circumstance that few of these recent papers have cited my invocation of methane in 1993. So I am a prophet with no disciples, and a wrong-headed prophet about the release of methane at that. In 1993 I accepted the conventional views that a major marine regression coincided with the mass extinction, and suggested that the drop in sea level could have released sufficient methane by removing the overlying pressure. The methane then caused global warming, given the sharp drop in carbon isotopes. Paul Wignall and Tony Hallam turned out to be right about sea level rising rather than falling, however, neatly eviscerating my argument.

There are several other possible sources of methane, so it could still be a contributing factor to the extinction. Methane could have been released from the deep continental shelves but only by considerable warming of surface oceans; as yet it is unclear how much warming would be required and whether a sufficient volume of methane would be available. Paul Wignall suggested that sustained Siberian volcanism induced a runaway catastrophic greenhouse effect that was eventually sufficient to release deep oceanic methane.[24] Evelyn Krull and Greg Retallack found multiple negative carbon shifts near the boundary in Australia and Antarctica.[25] Methane is a plausible explanation for the pattern they find, but only if the recharge rate for methane reservoirs is very high, higher than microbial ecologists find plausible. Methane release is undeniably an attractive hypothesis (aside from being a fad). It explains the carbon isotope shift, could induce global warming if a sufficiently large amount were released, and has the whiff of novelty. However, new estimates of the amount of methane in deep-sea sediments cast doubt on any of these ideas. I remain dubious that oceanic methane was the trigger of the mass extinction, partly because I doubt the volume was sufficient and partly because I think releasing a large volume of methane requires a mechanism that is more likely to have been the real trigger of the extinction. The one source that seems viable is volcanically induced heating of Siberian coals.

· · ·

So what did cause the greatest mass extinction in the past 600 million years, and perhaps the greatest in the history of life? The short answer is that we do not know, or at least I do not know. Several of my less reticent colleagues are sure they know but their answers are mutually contradictory and so cannot all be correct. We have growing evidence for an extraterrestrial impact, but evidence that is still, in my view, less than overwhelming; the evident coincidence of the Siberian flood basalts with the mass extinction, strong evidence for some degree of low oxygen and other changes in ocean chemistry, and a sudden spike in global temperatures. I have so often been wrong about the cause of the extinction that, in deference to my battered sense of scientific worth, I am tempted not to hazard an answer. But leaving the question hanging is hardly sporting, so here goes.

I am suspicious of coincidences, and as I write this, the eruption of the Siberian flood basalts seems a more plausible cause of the extinction than an extraterrestrial impact. While the simultaneity of eruption and extinction is impressive, the problem is how one caused the other. We have little understanding of the climatic effects of such massive eruptions and I wonder whether the progressive effects of acid rain produced by sulfuric aerosols, massive releases of carbon dioxide from volcanic eruptions and the destruction of coal beds with consequent cooling then warming, and possible impact of the thermal pulse at very high latitudes may have been far greater than we understand. It is within this nexus that I believe the cause of the extinction lies. Global warming may induce anoxia in shallow water simply because the oxygen solubility of water declines as temperature rises. I suspect that the shallow-water anoxia reflects the global greenhouse in the earliest Triassic. As to the deep sea, with evidence for pervasive anoxia, I think the evidence for continuing circulation is correct and that the anoxia lay within the sediments rather than the overlying waters. The chemistry of the deep ocean was odd, but was it odd enough to sequester the massive amounts of carbon dioxide and hydrogen sulfide, or methane, required by the Knoll oceanic overturn model? Probably not.

Impact enthusiasts claim that the simplest explanation is that an impact triggered the Siberian Flood basalt. That would certainly be an interesting result, and may be the only way the Permian will ever succeed in Hollywood, but nothing we know about either the Siberian volcanism or impacts provides much support.

Am I not dismissing the evidence for impact rather cavalierly? Perhaps, but geologists have been searching for, and rejecting, evidence of an end-Permian impact for twenty-four years now and even those favoring impact must acknowledge that the lack of iridium, reliable shocked quartz, or impact debris should counsel caution. It may well be that new evidence or new ways of studying data on hand will provide further support. What we need more than anything is confirmation of the hints of an impact from other sections and from independent laboratories. Ultimately such independent confirmation cemented the evidence for an impact at the Cretaceous-Tertiary boundary.

Our difficulties in understanding this event may lie more within how we tend to define a satisfactory answer than in the event itself. In the wake of the apparent success of the Alvarez-impact hypothesis many of us seem to prefer a single dramatic cause as an explanation for such events. Our knowledge of recorded history provides precious little support for such a view, and I see little reason, a priori to expect such a neat and tidy resolution to this riddle.

CHAPTER 9
Resurrection and Recovery

"This is the world's limit that we have come to; this is the Scythian country, an untrodden desolation." Aeschylus's description of the ravages of the Scythian hoards in *Prometheus Bound* is an equally apt description of the earliest Triassic. Triassic rocks are as barren of fossils as those of the precambrian. And just as the Guadalupian is the alternative name for the Middle Permian, geologists long ago christened the Early Triassic the Scythian, perhaps the most appropriately named interval of geologic time. The ancient civilizations in the Ukraine and central Asia survived the depredations of the Scythian hordes and life not only survived, but thrived after the end-Permian mass extinction. In that virtually complete desolation lay the resurrection of plant and animal life and eventually the diversity we see today.

Having made our way through the travails (both biologic and social) of the mass extinction, I can now confess that as an evolutionary biologist, understanding the recovery after the extinction poses a far greater intellectual challenge. Sadly few of my colleagues seem to share this perspective, and recoveries after mass extinctions have attracted far less attention than the extinctions themselves. In part this reflects a persistent bias that there is nothing particularly unique about biotic recoveries deserving of study: the extinction over, the survivors become fruitful and multiply, and that's that.

Pressed on the point, paleontologists and ecologists will extrapolate from what we know (or think we know) about the aftermath of a hurricane, widespread forest fires, or massive volcanic eruptions. Regrowth of a devastated area is spawned from survivors and immigrants from surrounding areas. Adjacent forests contribute windblown seeds and nearby reefs supply larvae bourn along by currents. To some extent the same processes occur after mass extinction, but new species are required to replace those that disappeared. The best analogy may be to oceanic islands, populated by chance migration of a few individuals from the mainland or other islands. The run-of-the-mill finches that first arrived on the Galapagos Islands from Ecuador found no real competitors for the diverse repast of seeds. Subsequent specialization on different seeds spawned many new species, the adaptive radiation that captivated Charles Darwin and generations of subsequent biologists. Much the same thing is thought to happen after a mass extinction.

So consider a chessboard, with each space on the chessboard representing a different species, each with its distinct ecological role (or niche). (We will assume that each species has a single distinct role, although often this is not strictly true.) Most models of mass extinction assume that extinction is a giant hand sweeping pieces off the board, but leaving the board and the rules of the game intact. Biotic recovery becomes a process of replacing the pieces on the chessboard and letting the game resume. As the survivors multiply into new species, the empty spaces on the chessboard become filled. A group of nearby spaces may be filled through the adaptive radiation of a single surviving species into a bush of similar forms. One of the rules of the game is that speciation is successful only when new species find an empty niche, symbolized as an empty space on the board. Once a niche is filled, the incumbent prevents a competitor from displacing it. (In computer simulations we can modify this rule so the challenger has some probability of displacing the incumbent, but we can ignore this for the moment. Computer simulations also wrap the board into a doughnut or torus so that there are no edge effects.) As the board begins to fill up again, there is less room for new species and speciation slows, beginning a transition to more normal evolutionary processes where the origination and extinction of species is more nearly balanced.

This analogy to a chessboard misses one critical feature: it fails to include the differences in the number of individuals (abundance) of different species, or of the varying numbers of species within different clades. This addition to the model is critical because some survivors become far more important than they were before the extinction, profiting by the demise of other groups.

In these final two chapters I will argue that this chessboard model has some fundamental problems, at least for understanding the aftermath of the end-Permian mass extinction. In my view, during some mass extinctions, the board collapses entirely, and as the game resumes, it becomes half chess and half backgammon, with some rules drawn from poker. Moreover, the rules evolve during the recovery: groups that initially are off to a promising new beginning may find themselves marching to their own demise as evolution adopts the rules of a new game. I find this view of the aftermath of extinction far more challenging, and far more revealing about the nature of evolution. But first we must begin with the events of the Early and Middle Triassic, the first 5–10 million after the mass extinction. The next chapter takes a look at the long-term implications of the end-Permian mass extinction for the course of evolution.

The poverty of latest Permian rocks in the United States is more than rectified by the abundance of Early Triassic deposits. The rising sea level that began in the latest Changhsingian continued through most of the 4 million or so years of the Early Triassic, laying down a thick blanket of marine rocks from Nevada through Utah and up into Montana and Wyoming. Such abundant and well-exposed rocks have made this one of the premier regions to explore the recovery. Hop out of a jeep among the scrubby piñon pine forests of southern Utah and you step onto pavements built of thousands upon thousands of specimens of the characteristic Early Triassic scallop *Claraia*. At Meishan in south China and along the Yangtze Gorge, *Claraia* is also found in astronomical numbers. Despite the incredible abundance of some species, the total number of Early Triassic species is a tiny fraction of those alive only a few million years earlier or later. Only twenty-two genera of fossils from the sea bottom have been recovered from Early Triassic rocks in the western United States—probably the lowest diversity for any 4-million-year interval since the precambrian, over 543 million years ago. Compare that with the three hundred or so species from Meishan during the last few hundred thousand years of the Changhsingian, or the thousands of species from the Capitanian reefs of west Texas.

These tens of thousands of *Claraia* are weeds, the ecological equivalent of dandelions springing up unbidden in a spring lawn. Such opportunists serve an important ecological role, taking advantage of disturbance to colonize new areas as quickly as possible and with an overwhelming proliferation of individuals. Just as the aftermath of forest fires brings forth ferns, grasses, and quick-growing trees, and a good hurricane clears away the slower growing corals on a reef to allow others to spread, mass extinctions call forth opportunists as well. Most of the fossils in the first 3 million years or so of the Early Triassic share the ability to rapidly produce large populations.

These opportunists do not quickly disappear, giving way to longer-lasting and more diverse assemblages. For most taxa these low-diversity assemblages persist through the Early Triassic. Ammonoids, conodonts, and a few other groups are exceptions, but the opportunists of the earliest Triassic are not displaced by a surge of new species and the appearance of diverse ecosystems. Instead, there is a prolonged period of the blahs. Some new species appear, and some have argued there was no long hiatus before recovery. The point, however, is not that a few new species appear, but that ecosystems remain depauperate for several million years.[1]

Not until the Spathian, the last of the four substages of the Early Triassic (figure 9.1) and 4 million years after the mass extinction is there a demonstrable increase in speciation, and the onset of what we can truly regard as a recovery. Compare this to each of the other five mass extinctions when either there was no survival interval or it was very brief, perhaps a hundred thousand years at most. Although we still lack radiometric dates for the duration of the Griesbachian, the presence of several conodont zones and the thickness of rock in many areas suggest it may have lasted more

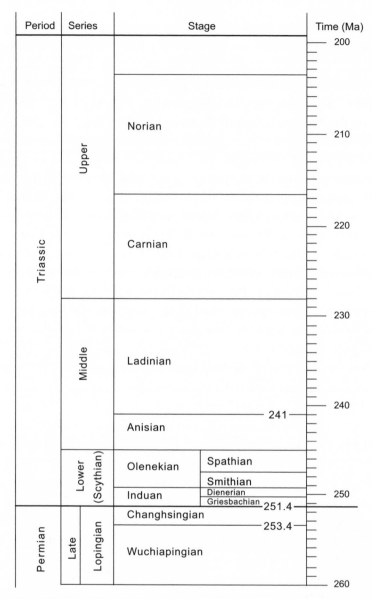

Figure 9.1 Triassic timescale. The survival interval for marine ecosystems is from the Greisbachian through the Smithian, with the real biotic rebound in the Spathian and Anisian. The Scythian is the informal geologic name for the Early Triassic.

than a million years. The unprecedented length of this survival interval is the first of two conundrums of the recovery.

The second, and perhaps greater conundrum is the resurrection of genera that had been missing from the fossil record for millions of years. This sounds odd, of course, since species that go extinct cannot really be resurrected. Roger Batten of the American Museum of Natural History in New York first called attention to this phenomenon in a 1973 paper with the wonderful title: "The vicissitudes of the gastropods during Guadalupian-Ladinian time." Roger noted that representatives of thirty-two genera and sixteen families of gastropods were found in Middle Permian and Middle Triassic rocks, but had never been recorded from the intervening rocks of the latest Permian and the Early Triassic. This curious phenomenon later received the appellation "Lazarus taxa" by David Jablonski for groups that appeared to have gone extinct, but then reappeared. Since 1973 the Lazarus phenomenon has been recognized among brachiopods, echinoids, bivalves, and other groups. I have been tracking the gastropods, and about 30% of all genera were Lazarus taxa during some point between the mid-Permian and mid-Triassic.[2]

The abundance of these Lazarus taxa is revealing. Obviously the fidelity of the fossil record was not high, but were these missing species hiding in small refuges, not preserved for some reason, or have we have somehow misinterpreted the fossils? Almost as curious as the disappearance of so many lineages is their rapid reemergence during near the end of the Early Triassic and the beginning of the Middle Triassic, coincident with the appearance of many new lineages.

The lengthy survival and recovery interval after the end-Permian mass extinction has been evident for years yet the cause remains frustratingly obscure. Several explanations have been proposed; none is wholly satisfactory. The long lag could simply reflect the magnitude of the extinction. Since the end-Permian mass extinction was greater than any other mass extinction of the past 600 million years, we might expect the length of recovery to scale with the magnitude of the extinction. In essence, the ecological and evolutionary disruption caused by the extinction would require the gradual rebuilding of primary productivity, followed by

expansion of higher levels of the food web. Plants and other photosynthetic groups toward the bottom of the food chain should
begin to diversify first, with predators the last to recover.

While this theory of ecological retardation sounds convincing,
there is no evidence that the duration of survival and recovery
intervals scales to the magnitude of the extinction, although with
only five great mass extinctions there really is not enough data for
a convincing test. Although we lack reliable radiometric dates for
most postextinction rebounds the available evidence shows no relationship between extinction magnitude and duration of recovery. Also, the return of so many Lazarus taxa coincident with the
appearance of so many new species remains unexplained. Why
should these have happened at essentially the same time?

Paleontologists tend to make the assumption that the conditions that produced the extinction magically ameliorate after the
peak of the extinction. There is no particular reason why this
should be so. The number of apparent extinctions may decline
simply because everything that is likely to go extinct has gone extinct. The ecological retardation hypothesis assumes that the environment returned to normal with the conclusion of the mass extinction, in the sense that there was oxygen in the ocean, climate
was not unusually harsh or unstable, and there was no acid rain
or other physical challenges.

In 1991 Tony Hallam, one of the chief purveyors of the anoxia
model for the extinction, suggested that low-oxygen waters persisted through much of the Early Triassic and retarded the recovery. Several other extinction triggers could also persist for hundreds of thousands to several million years, retarding any biotic
recovery. Continuation of Siberian flood basalt volcanism, as an
example, may have hampered recovery through production of acid
rain and climatic fluctuations. This does not mean that new species
would not appear during this interval, far from it. The persistence
of life today under incredibly harsh conditions reminds us than an
environmental challenge is an evolutionary opportunity. Diversity
in newly challenging circumstances may be low, but natural selection is sufficiently powerful that the survival interval may represent
nothing more than a prolonged time where harsh environmental
conditions impeded the recovery of most groups.

We have two clear alternative explanations for the pattern of biotic recovery in the aftermath of the extinction: environmental retardation caused by persistent adverse environmental conditions or simply a very long delay in ecological recovery induced by the magnitude of the extinction. A look at the record of different groups and different environments may help test these different perspectives.

One of the Early Triassic opportunists suggests just how far the evolutionary clock was reset by the extinction. The earth's first recorded biodiversity crisis was the decline in microbial mats spurred by the spread of burrowing animals and changes in ocean chemistry some 600 million years ago. We met stromatolites earlier, the thin layers of sediment commonly built by complex microbial communities. Photosynthesizing microbes form the upper surface of a stromatolitic mat. Bacteria favoring anaerobic or otherwise unsavory chemical environments find more amenable housing farther down in the mat. This thin veneer of living microbes caps layer upon layer of dead cells and trapped sediment, with the lower layers slowly turning to rock. Many stromatolites are simple mats, but microbial communities can generate a variety of domes, branching columns, and other structures. Just east of Death Valley in California is a huge mushroom of rock, probably fifty feet high and over one hundred feet across the base. Six hundred million years ago this was a single stromatolite living in some quiet lagoon. Stromatolites still persist in the Bahamas, Baja California, and western Australia but are far less common than they were 1 or 2 billions years ago. Curiously, however, time reversed with the mass extinction, and stromatolites are widespread in earliest Triassic oceans, reclaiming habitat they last occupied 400 million years earlier.

Ask a paleontologist about stromatolites and he or she will immediately describe the primitive seas of the precambrian. But the term *stromatolite* was first applied by the Polish geologist Ernst Kalkowsky to structures in the Early Triassic Buntsandstein in the Harz Mountains of central Germany. These stromatolites grew in a playa lake rather than the ocean, but other Early Triassic stromatolites have been found in marine rocks in the western United States, Turkey, Greenland, south China, Iran, and Japan. In some of these localities, particularly areas in Nevada and Utah studied by Sara Pruss, a graduate student at the University of Southern California, the microbial textures occur, at least episodically, through the Early Triassic.[3]

Microbes also form structures known as wrinkle marks (the original German term is more euphonious: *Runzelmarken*). Whether one calls them opportunists, forms that occur sporadically as fossils but flourished in the aftermath of even small disturbances; disaster species, taxa that are rarely seen as fossils and are called forth only in the wake of great disasters; or simply anachronistic elements of an earlier world, the proliferation of microbial-dominated environments through the Early Triassic is a testament to its unusual nature. The clock in near-shore environments was reset to the late precambrian for 4–5 million years, based on the record of wrinkle marks in the late Spathian Virgin Limestone in Nevada.[4]

The inarticulate brachiopod *Lingula* is often described as a "living fossil" because it has persisted seemingly unchanged for hundreds of millions of years. Since *Lingula* basically looks like a watermelon seed, there is a limit to how much evolutionary change one could detect. But its ecological behavior is enlightening. *Lingula* is another of the Early Triassic disaster species and one of those twenty-two genera found in the western United States.[5] Although *Lingula* is commonly restricted to low oxygen (dysaerobic) and near-shore areas, in Early Triassic rocks of the intermountain West it occurs across many environments. As sea level rose during the Early Triassic, the Dinwoody Formation was deposited on a shallow shelf occasionally swept by storms. *Lingula* was the dominant fossil in three of the four distinct environments of the Dinwoody, ranging from near-shore muds to silt and limestones farther offshore, and comprised 49% of all the fossils. In contrast to the normal restriction of *Lingula* to unfavorable environments, when the Dinwoody Formation was formed it was abundant across a normal marine shelf in a diverse range of environments. This burst of *Lingula* is found in many other places in the world too. Later in the Early Triassic *Lingula* became rare.

Claraia, stromatolites, and *Lingula* all thrived during the first few hundred thousand years following the extinction. Fossil faunas in

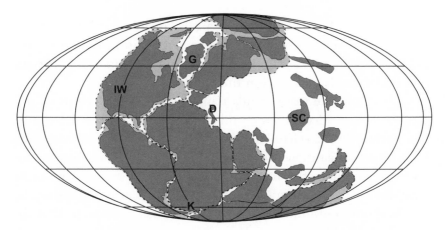

Figure 9.2 Paleogeographic reconstruction of the Early Triassic showing the position of significant localities: D: Dolomites (northern Italy); G: Greenland; K: Karoo (South Africa); IW: Idaho/Wyoming; SC: south China.

Utah, the Alps, Iran, Pakistan, and elsewhere (figure 9.2) all show this burst of extraordinarily abundant opportunistic survivors with a dearth of new species and depauperate ecological communities characterizing the immediate postextinction or survival interval. After their burst of glory, each lineage retreated into its previous obscurity without contributing much to the later biotic recovery.

The Dolomite Alps of northern Italy have a very different kind of grandeur from the dry desert of Utah. Beautiful, picturesque, and clogged with summer tourists, the dolomites are built of Triassic limestones with the lower portions formed at the same time as the Early Triassic of Utah. There is a much greater range of environments than in Utah, from near shore to the edge of the continental shelf. As in the Early Triassic of China, thinly laminated lime muds populated with a very low-diversity fauna occur through the Griesbachian. Richard Twitchett, who we first encountered at the end of chapter 6, spent his graduate research unraveling the patterns of recovery in the Werfen Formation.[6] The sediments, lack of burrows, limited number of fossils, and geochemistry all suggest these laminated layers were deposited under low oxygen conditions that persisted after the mass extinction. By the late Griesbachian the abundance and diversity of burrows increases but shelly fossils remain few and far between. This suggests that while the environmental stress of the early Griesbachian had ameliorated, it had hardly disappeared. The initial recovery of the late Griesbachian seems to have reversed again in the Smithian, before a strong increase in the diversity and abundance of burrows and shelly fossils occurred in the Spathian.

Fossils from the wonderfully named Wadi Wasit locality in the Central Oman Mountains confirm the significance of anoxia in the Early Triassic (Griesbachian). The earliest Triassic deposits have opportunistic bivalves much like those in Utah, but by the late Griesbachian, the fossils include a diverse assemblage of at least twenty-five genera of bivalves, small gastropods, crinoids, brachiopods, ammonoids, echinoids, and ostracods in 6–12-inch-thick beds. Unlike the Alps, the environment was well oxygenated, and the presence of crinoids and echinoids suggests the absence of other environmental stresses such as abnormal salinity. Richard Twitchett has argued that the well-oxygenated nature of this deposit, in contrast to communities in Utah and Italy where at least occasional evidence of anoxia has been found, supports suggestions that persistent anoxia through the Early Triassic retarded recovery in most regions.[7]

The diversity of fossils at the earliest Triassic Wadi Wasit locality, deposited in an apparently well-oxygenated setting, is the best test of Hallam's suggestion that persistent anoxia explains the delayed Early Triassic recovery. However, since atmospheric oxygen is rapidly mixed into the shallow ocean by wave and current activity, persistent anoxia over millions of years is only plausible if atmospheric oxygen levels were drastically reduced. As I pointed out in the previous chapter, oxygen levels in water fall as temperature rises, so high global temperatures could cause apparent anoxia in seawater.

Early Triassic rocks in the Death Valley region of California complicate the picture. The Union Wash Formation was deposited on the outer continental shelf at the Smithian-Spathian boundary in deeper water than the Dolomite Alps. Anoxic sediments, carbonate cements on the seafloor, and the dissolution of ammonoid shells before burial all suggest fairly harsh conditions. The carbonate cements are similar to those described by John Grotzinger as formed by the mixing of a deep, alkaline-anoxic water mass with shallow, oxygenated water.[8] The unresolved question is whether such a harsh environment persisted from the mass extinction, or

alternated with more benign conditions through the Early Triassic, allowing a greater diversity of organisms to thrive. These Early
Triassic vignettes remain too few to provide a convincing global
picture of events.

Conodonts and ammonoids are conspicuous exceptions to this
generalization of delayed recovery, but in ways that may support
the general conclusions about the significance of environmental
effects. Conodonts suffer little extinction at the Permo-Triassic
boundary, so their continued speciation through the Early Triassic
is perhaps not surprising, but ammonoids experienced catastrophic extinction. Thus the rapid proliferation of new Early Triassic ammonoid species is somewhat unexpected. Although the magnitude differs depending on the diversity metrics used, ammonoids
diversified during the Early Triassic, but the forms of these new
species exhibit a more complicated pattern. If the Permian extinctions had opened up new ecological space for ammonoids, then we
expect great morphological variability through the Early Triassic.
Some surviving ammonoids with extreme morphology died out in
the Griesbachian, but new Dienerian ammonoids were more similar to the norm, leaving Dienerian ammonoids with less morphologic variability than their Griesbachian ancestors. This "missing
variability" among Early Triassic ammonoids likely reflects continuing environmental stress, restricting ammonoids to swimming high
in the water column and limiting bottom-dwelling forms.[9]

The Sinbad Limestone of the Moenkopi Formation of Utah
dates from the middle of the Early Triassic, slightly younger than
the Griesbachian. Of the sixteen genera and twenty-seven species
of snails, almost all are very small individuals, mostly less than one
centimeter, and while some may represent juvenile forms, most
were adults. Larger gastropods are virtually absent. Microgastropods are known from other intervals, although how they are is
unclear because most collecting strategies would miss them.[10]

But is the dominance of microgastropods in Utah indicative of
conditions in other parts of the world? Jonathan Payne, recently
a graduate student with Andy Knoll at Harvard, set out to evaluate
this by collecting global data on the size of snails through the
Permo-Triassic extinction and into the Early Triassic. His results
show that the pattern in Utah is typical. Five–10-centimeter snails

are typical of the Permian and Middle Triassic, but he could find no record of any Early Triassic snails larger than 2.6 centimeters. There may have been some preferential extinction of larger-sized snails during the extinction, but the limited number of species and their small size may account for much of the Lazarus phenomenon among the group. Nonetheless, the abundance of these microgastropods in Utah and the broad geographic extent of such small fossils through the Early Triassic suggests they are at least plausibly related to continuing environmental pressures, since other microgastropod faunas are also related to stressed environments. As Payne points out, however, sources of stress could include changes in food supply or primary productivity, predation pressure, oxygen availability, or other factors.[11]

$\cdots$

Modern reefs of scleractinian corals, abundant fish, and myriad other creatures filling every crevice were born in the wake of the end-Permian mass extinction. The tabulate and rugose corals of the Paleozoic had disappeared and it was not until the distantly related scleractinians appeared in the Middle Triassic that reefs really began to recover. Early Triassic reefs did exist, but they owed greater allegiance to the precambrian than to the Paleozoic, for microbes largely built them. In the Nanpanjiang Basin of the southern China province of Guizhou, just north of Vietnam, earliest Triassic reef buildups are up to fifteen meters thick. These mounds were far smaller than what we think of as a reef today (in fact geologists generally call them mounds and biostromes to distinguish them from the larger structures), but they functioned in much the same way as modern reefs. In Guizhou, the most extensive mounds are found in earliest Triassic rocks. Later in the Early Triassic less extensive mounds were embedded in shallow water limestones. Other fossils occur in the reef, including gastropods, bivalves, ostracods, worm tubes, brachiopods, and some foraminfera, but in keeping with the time, diversity was very low. Similar Early Triassic microbial reefs were widespread in southern Turkey, Armenia, Iran, and southwestern Japan, and increased during the late Early Triassic.[12]

The earliest scleractinian corals are found on the Middle Triassic carbonate platform of Guizhou and had already diversified into several different forms. The ancestry of these corals is disputed because of the long gap between the last Permian rugose and tabulate corals and because the origins of the scleractinians are obscure. The construction of scleractinians differs considerably from the Paleozoic corals, but an increasing number of unusual "scleractinomorph" (scleractinian-like) corals have been discovered in the Paleozoic. These fossils could be ancestral to the scleractinian corals, but George Stanley has recently suggested an alternative: that scleractinian-like corals arose multiple times over the past 540 million years from unskeletonized sea anemones (evolutionary convergence again). Molecular studies of living corals lend support to the hypothesis that the scleractinians originated several times. In other words, the scleractinians may not be a single clade, but a group with very similar morphology and several ancestors.[13]

The multiple origins of scleractinian corals during the Middle Triassic has some interesting implications, for it suggests that "naked corals," as Stanley puts it, acquired skeletons repeatedly but only became evolutionarily and ecologically successful in the aftermath of the mass extinction. Rather than the diversification of a single group driven by the opportunities of the recovery, akin to the adaptive radiation of Darwin's finches, the situation is more complex. The molecular data, albeit with considerable uncertainty, suggests that some of the scleractinian ancestors arose well before the mass extinction, and the fossil record suggests that such innovations had been occurring steadily during the Paleozoic. They were not successful until the Middle Triassic, when several groups each established themselves.

Scleractinians were not the only major group of organisms to appear during the Anisian. After the microbially dominated reefs of the Early Triassic, sponges, algae, and clams all became major constructors of reefs, which almost reached the diversity of Late Permian reefs. Clearly something more than simply the appearance of scleractinians was responsible for the spread of reefs during the Middle Triassic. One possibility is that the environment ameliorated to the point where all of these groups could increase in abundance sufficiently to produce the massive biologically constructed features we recognize as reefs. Recall from chapter 5 Rachel Wood's argument that reefs require broad carbonate shelves. So the physical setting needed to form reefs might simply not have existed during the Early Triassic, thus retarding the earlier appearance of reefs. The spread of Middle Triassic reefs does coincide with the expansion of carbonate platforms in the Mediterranean, Oman, Pakistan and India, and China, although some carbonate platforms existed in these areas in the Early Triassic.[14]

Reef recovery started later than that of other groups, and even during the Anisian reef, biotas were not particularly diverse. Lazarus taxa among reef organisms reappear later than in other ecosystems. For example, some sponges and other organisms did not reemerge until the Norian of the Upper Triassic. A major unresolved issue was whether the re-expansion of reefs was limited by the evolution of groups adapted to reefal environments, or whether the physical setting was missing. If the latter, reefs could have made an earlier reappearance.

. . .

The pervasive extinction of plants and the destruction of land ecosystems in the latest Permian produced an entirely different flora in the Early Triassic. Early Triassic landscapes in the Sydney Basin of Australia had only a few species: small, spiky bunches of quillworts, a few shrubby lycopods, a seed fern, horsetails, and a few conifers. Quillworts, including the Early Triassic genus *Isoetes*, are alive today and look like small clumps of stiff grass, but with hollow leaves and roots, hence the name quillwort. The earliest Triassic species *Isoetes beestonii* had fleshy leaves like a modern succulent and its abundance seems to suggest that large swaths of *Isoetes* lived along the margins of lakes or ponds. *Isoetes* was a weedy opportunist, and thus its success in the earliest Triassic is not surprising. During the Early Triassic, at least five new genera and thirteen species appeared, some forming tall stalks and other minute, spiky cones. It spread from semi-aquatic habitats into areas as diverse as salt marshes, mangrove swamps, flood plains, and even arid deserts.[15] In contrast to the multiple originations of the scleractinian corals, the quillworts do seem to have undergone a classic adaptive radiation.

The success of these weedy Early Triassic quillworts is a bit surprising. The earliest Triassic *Isoetes* were well adapted for colonizing disturbed areas, and while the later species may have been well adapted to the new environments they occupied, the clade as a whole had generally poor competitive ability compared to other clades. Unless disturbance is relatively frequent, most weeds give way to more competitive shrubs and trees within a few tens of years. That *Isoetes* was successful enough to spread into so many different habitats illustrates how few other plants were present. But *Isoetes* was only a part of the low-diversity Early Triassic floras. In addition to quillworts, small conifers, a few shrubby lycopsids are widespread. These same plants are found in many areas of the world, so as with the oceans, the land was dominated by a few, widespread species.

Quillworts spread across earliest Triassic tropical landscapes of Europe and with lycopsids composed the bulk of the flora until the Middle Triassic when conifers finally reclaimed their position as the dominant group of plants. Cindy Looy's graduate work at Utrecht University in the Netherlands involved detailed analysis of the pollen record through the Early Triassic across Europe.[16] Pollen does not reveal what plants were at a specific site as precisely as fossil leaves or other plants parts, but it provides an unparalleled record of regional plant distributions. At site after site Cindy found a very similar pattern, with lycopsids thriving immediately after the extinction. A large, succulent quillwort, *Pleuromei sternbergii*, was particularly common, as were other shrubs. But the poorly developed soils of the Early Triassic show little sign of trees. Finally, about halfway through the Spathian stage, conifer-dominated floras swept across Europe, marking the end of the survival phase and the onset of true recovery. This pattern of change in plant communities is very similar to the response to the end of the ice age 11,000 years ago, with the early herbs and shrubs gradually being replaced by conifer forests. The difference, however, is the need for the evolution of new species in the Early Triassic.

In north China the same pattern appears: Early Triassic plants include large numbers of *Pleuromeia* and *Isoetes*, ferns and other plants similar to those in western Europe. Unlike western Europe, however, *Isoetes* and other lycopsids persist into the Middle Triassic. The rebuilding of the north China flora was a gradual expansion and there is no sudden return of conifer forests. Vegetation slowly returned to waterways, horsetail marshes gradually covered wetlands, and other plants adapted to the seasonal fluctuations between dry and wet. Not until late in the Middle Triassic did widespread vegetation reappear.[17]

In chapter 6 I described the disappearance of coals from the low-oxygen, acidic environments needed to form peat and, with time, coal. This striking lacuna continues through the Early Triassic. The glossopterids, lycopsids, cordiates, and ferns that produced Permian coals are missing. Middle Triassic coals are thin and rare, and the earliest examples are coincident with the increased diversity of fossil leaves. In the Anisian of Australia and New Zealand, the seed fern *Dicroidium* formed thin coals along with some cycadophytes and ferns. Not until the Late Triassic do other plant groups began to move into peat swamps and coals again become prominent.[18]

The coal gap might reflect a lack of environments to produce coals, or suddenly more effective herbivores that ate the plants, or increased efficiency of the fungi responsible for decomposing plants. There is good evidence for global warming in the earliest Triassic, and this may have reduced the number of places where coals could have formed. But appropriate settings with sufficient water did exist and there is no evidence that increased herbivory or faster decay prevented coals from accumulating. The dearth of coals through the Early Triassic and their scarcity in the Middle Triassic reflects the difficulty of new plants in adapting to the demanding chemical environment of peat mires. For coals to form again competition between plants in other lowland habitats had to become high enough that adaptation to a swamp became a reasonable option.

The high-latitude floras of Australia and Antarctica, the equatorial landscapes of Europe and the distribution of coal all document a lengthy survival interval through much of the Early Triassic, and a very rapid reappearance of conifer-dominated forests near the beginning of the Middle Triassic. The first coals develop at the same time, although thicker coal seams evidently required the evolution of plants adapted to the specific environments of peat mires.

Early Triassic soils in Australia and Antarctica and river-drainage patterns and sediment deposition in South Africa and Russia all suggest a catastrophic loss of vegetation and sudden global warming at the Permo-Triassic boundary. In Australia and Antarctica, for example, evergreen conifers in a cool, humid setting with nutrient-poor soils replaced the humid, cold swamps and lowlands with broadleaf deciduous trees. Since the latitude of both places did not change suddenly, the shift in flora and soils must indicate a much warmer climate. The berthierine in earliest Triassic soils from Antarctica suggests low oxygen settings, possibly from methane release (chapter 7).[19] It is unclear how long these warmer climates persisted. Neither coals nor deciduous, broadleaf forests return to Australia and Antarctica until the Middle Triassic.

Early Triassic vertebrate assemblages follow a similar pattern to land plants and marine invertebrates: relatively few species but these species were broadly distributed. The Early Triassic dicynodont *Lystrosaurus* was a barrel-chested therapsid about the size of a large dog, with a shoved-in snout and two large tusks, homely enough to make a bulldog look attractive. But give *Lystrosaurus* credit, for it was the predominant Early Triassic vertebrate from South Africa, to India, Antarctica, China, and Russia. In their work in the southern Ural Mountains, Benton and his Russian colleagues found that Early Triassic vertebrate assemblages had few species and showed none of the increase in diversity seen after earlier Permian biotic turnovers. By the end of the Middle Triassic, more than 15 million years after the extinction, big herbivores and top carnivores as well as small fish-eating tetrapods and small insectivores were still missing from the fauna, suggesting that vertebrate recovery was particularly delayed.[20] Later in the Early Triassic *Lystrosaurus* was succeeded by other therapsids that also had wide geographic distribution.

· · ·

Detailed carbon isotope curves of the Early Triassic are now available from Jonathan Payne's work in the Early Triassic of Guizhou, in southern China where he found that the major carbon shift at the Permo-Triassic boundary was the first of three major changes in the carbon cycle. Additional negative swings occurred at the end of the Griesbachian, in the late Smithian and near the end of the Dienerian, with a major positive swing near the Dienerian-Smithian boundary. In the Middle Triassic the carbon record settles down to steady values of $2^0/00$. The persistence of these major shifts in the carbon cycle through the Early Triassic may reflect environmental disturbances, or simply decimated ecosystems.[21]

As always, we must carefully examine the quality of the data we are using, and difficulties in both preservation and identification may exacerbate the apparent magnitude of the delayed recovery. Not all Lazarus taxa are quite what they seem. A decade ago sponges and many other groups found on reefs appeared to be good candidates for Lazarus taxa. Very similar forms had been recovered from the mid-Permian and mid-Triassic, but upon closer inspection Erik Flügel realized that the Permian and Triassic species were not actually as closely related as they first appeared: the similarities were due to convergent evolution to a similar form, not shared descent from a common ancestor. Convergence is particularly common if the number of engineering solutions to a particular problem is limited. In such circumstances very similar-looking species may appear in different clades, or at very different times. The term Lazarus taxa had been such a hit that my friend Mary Droser, a paleontologist at the University of California, Riverside, and I decided that these convergent forms needed a good name and somehow "Elvis taxa" was born, for the many Elvis impersonators spawned since the death of the King. (Dave Jablonski, otherwise a good friend of mine, was so annoyed by our apparent trivialization of his Lazarus taxa that he still refuses to acknowledge our seminal contribution to paleontological thought.) Very careful work is required by specialists in the morphology and systematics of these groups to distinguish true Lazarus taxa from Elvis taxa, but they tell us very different things about the extinction and recovery. Lazarus taxa reveal how poor the fossil record is while Elvis taxa suggest extinction was greater, but the range of possible forms is more limited than we might otherwise suspect.

In addition, the quality of many Early Triassic fossils is really pretty lousy. Graduate students often do not want to work on the Early Triassic because there are so few species that fieldwork quickly becomes boring. With the exception of some ammonoids, most fossils are poorly preserved and difficult to identify. Paleontologists tend to assign poorly preserved fossils to common, widely distributed groups, although species of many different genera may look very similar. Among Jennifer Schubert's gastropod fossils from the Virgin and Thaynes members of the western United States, *Worthenia* is a proxy for a host of squat, turreted snails, while *Zygopleura* and *Coelostylina* may encompass a host of high-spired forms. The problem is not that I did a bad job when I helped identify these fossils in 1991, but that the material is simply too poorly preserved for more definitive judgment. Sometimes the most appropriate comment in science is "I don't know."

Along the Yangtze Gorges in China you can pick out the Permo-Triassic boundary from the middle of the river: it is the point where the reddish-brown blobs of chert disappear. The grayish-black limestones below are full of chert, but above the boundary, the chert has vanished. Recall that the most exquisitely preserved fossils from the Middle and Late Permian in west Texas, China, Pakistan, and elsewhere are commonly silicified. In the west Texas Permian Basin faunas, George Girty described 113 different species of brachiopod, in his 1908 monograph on the Guadalupian fauna, none of them silicified. Over more than three decades of fieldwork, G. Arthur Cooper and Dick Grant found some nine hundred silicified brachiopod species. Coop's determination to collect every last brachiopod species was such that he and Dick suffered through years of enervating summer fieldwork before Dick finally pointed out that unlike their colleagues at universities, as Smithsonian employees, they did not have to do fieldwork during the summer. As Dick recounted the tale, Coop thought about that for a while . . . and the next year they went to Texas in April. Although Cooper and Grant's work lasted longer and covered a larger region than Girty's, the chief reason they found eight hundred additional species was the silicification. Most of these species and many genera are known only from silicified specimens. No silica, no specimens. The silica came from siliceous sponges, and when they largely disappeared during the mass extinction, silicification did as well. During the Early Triassic very few fossils were silicified and so we simply cannot say when species normally preserved only as silicified forms disappeared, or how abundant they may have been. Both poor preservation and the absence of silicification probably magnify the apparent lag in recovery, but it is notoriously difficult to test for the lack of something, so the significance of these factors is hard to determine.[22] There is no a priori reason why the preservational problems that affect marine fossils should also affect plant and vertebrate fossils on land, and the similarities of the marine and terrestrial patterns suggest that while preservation may have magnified the delay in recovery, the pattern is real and requires explanation.

The delayed recovery among most marine groups is mirrored by the long lag before diversification begins among plant groups, and corresponds to an interval with numerous very large shifts in the carbon cycle. The Lazarus taxa return near the end of the Triassic, close to the end of the isotopic shift and the return of diverse marine communities. All of this strongly suggests that the primary cause for the delayed Early Triassic recovery was continuing environmental disturbance that retarded the possibility for biotic recovery. The source of this continuing disturbance is unclear, and the timing of the Siberian traps makes volcanism an unlikely cause.

· · ·

Let's return to the simple chessboard model I introduced at the beginning of this chapter, where each square represented the ecological requirements of a single species, and extinction eliminated pieces from the board, but left the board and the basic rules of play essentially intact. Those empty spaces on the board represent niches vacated by extinct species, available to be refilled after the extinction, and they are the essential ingredient of this model. If we track the progress of recovery, we anticipate an initial burst of new species as the survivors proliferate into the vacant opportunities (figure 9.3). With the rule that once a space is filled, incumbents are likely to resist displacement by a new species, the rate of successful new species should fall as the spaces on the board fill

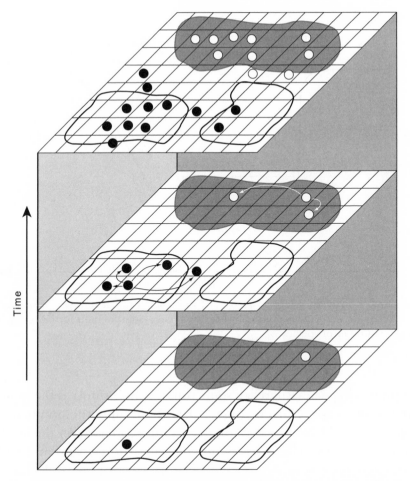

Figure 9.3 Chessboard model for postextinction biotic recovery. The earliest interval is on the bottom, and each square denotes a different potential niche; shaded areas show general adaptive zones, such as the general ecological role of clams or crabs. New species appear through time, with most species evolving into a niche adjacent to the ancestral species, but with some making longer evolutionary and ecological jumps. As the space begins to fill, the rate of successful new speciations declines. But such models depend on niches existing independent of the species that fill them, instead of being constructed by the species.

up. This conceptual model, as well as mathematical models derived from ecology and island biogeography (chapter 2), suggests that the recovery should follow a sigmoidal pattern with slow diversification followed by an exponential rise in the number of species and then a decline in the number of new species (so the total

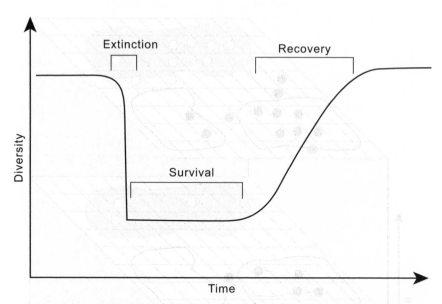

Figure 9.4 Pattern of logistic recovery from a small initial number of species, with the rate of recovery eventually leveling off as resources limit further growth in the number of species.

number of species will reach some rough equilibrium) as the number of open spaces drops (figure 9.4). With time, the number of species should reach some rough equilibrium based on the available resources. The almost imperceptible initial response will probably appear in the fossil record as a lag or survival phase, with the exponential increase evident as the onset of the recovery phase. Since new species are almost by definition not very abundant, they will rarely show up in the fossil record. New species will only appear in the fossil record when they become abundant. Such models generally suggest that the duration of the lag should be proportional to the magnitude of the extinction: the larger the extinction, the longer the lag.

The complexities of actual recoveries will produce patterns that vary considerably from these expectations. The faster generation times of some microfossils such as forams allow more rapid evolution than the slower turnover of land-dwelling vertebrates, for example. Likewise, environmental factors may favor recovery in one part of the world over another, producing geographic variations in the pace of recovery.

Two peculiarities of how paleontologists have approached these issues may have exacerbated our problems in understanding biotic response to mass extinctions. Many paleontologists persist in using the number of species (or higher taxa) present as a metric of recovery. But the response to a mass extinction is above all an ecological one, and the number of taxa is only an indirect measure of evolutionary activity. Understanding these events requires an ecological approach. What are the dynamics and interactions between species? How well were ecosystems functioning? How complex were food webs and other ecological services during the survival and recovery intervals? Perhaps most importantly, how does the structure of the ecosystem facilitate the speciation involved in biotic recovery? Simple models do not address these issues.

Beyond this, I believe our thinking has been blinkered by over-reliance on ecological models of recovery derived from vastly smaller disturbances. Cindy Looy noted that the patterns of Early Triassic floras in Europe are very similar to the return of forests after the end of the Pleistocene glaciations, but the time scale is different. Indeed the very words we use, *recovery* and *rebound*, reflect this reliance upon ecological models that may be wholly inappropriate. Rachel Wood and I have spent many a bottle of wine trying to come up with better terms than these, but to no avail, so deeply, and misleadingly, are they embedded in the way we think.

We must develop models tailored to the ecological and evolutionary processes of recovery after mass extinction, explore the kinds of data paleontologists would need to collect to test each of the possibilities, and then design fieldwork explicitly to produce the information required. With Ricard Solé of the University of Barcelona, we are following just this approach. Two of the models are worth discussing briefly, the first because of how much is revealed by a relatively simple approach, the second because I think it is closer to the processes actually driving biotic recovery. The first model uses a simple ecosystem with plants (or some other primary producer), herbivores, and carnivores. There are links

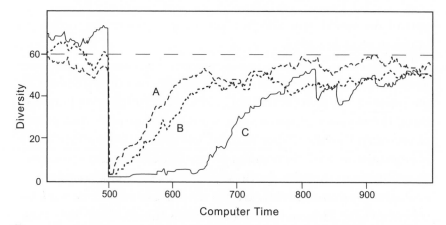

Figure 9.5 Variation in recovery patterns between three different ecological levels with time: primary producers (A), herbivores (B), and carnivores (C). This is a very simple model, but it shows the importance of considering different trophic levels. After Solé et al. (2001).

between various species, with herbivores feeding on plants, and carnivores feeding on herbivores. Extinction is imposed simply by eliminating some fraction of the primary producers, which has the effect of quickly killing off herbivores and carnivores linked to the primary producers. We have rules for creating new species at each of the three levels, and for establishing the connections between species.[23]

Computer simulations of recovery based on this model produce successive waves of new species at each level: plants respond relatively quickly, herbivores more slowly, and carnivores yet more slowly (figure 9.5). This provides us with patterns that we can test in the fossil record: can we detect more rapid response among primary producers than for species at the top of the food chain? Additional simulations in which the magnitude of the mass extinction varies between simulations produces another intriguing pattern: recovery is very rapid, with no relationship between extinction magnitude and duration of the recovery for small extinctions, but eventually a threshold is reached beyond which the length of the recovery scales with the size of the extinction. How these thresholds are produced remains unclear, but they suggest that the interconnections between species within an ecosystem may strongly influence the course of the recovery. Since paleontologists have long expected that there should be some proportionality between the magnitude of the mass extinction and the duration of the biotic recoveries, this is not a particularly surprising result. The absence of evidence for this proportionality may simply mean that the data we have are not good enough to recognize it. In any event, these results are clearly different from the simple chessboard model.

If the length of a recovery is scaled to the size of the extinction for larger mass extinction but not to smaller biotic crises, this may indicate that there is a point where the ecological fabric is irreparably destroyed. Below this threshold enough biodiversity may be preserved that recovery may truly be a process of refilling empty niches as in the chessboard model. Construction of these ecological networks is a wholly different issue than repairing some missing connections. This leads us to the second model, which focuses on the construction of these ecological relationships, rather than the refilling of them.

One of the fundamental underpinnings of the chessboard model is that niches exist independent of the species that fill them. Niche theory has a complicated history in ecology, with two distinct schools dating back to the 1920s that differ significantly regarding the extent of an organism's contribution to the dimensions of a niche.[24] While I am willing to accept that resources may exist for new species to exploit, I reject the idea that empty niches exist independent of an actual species to fill the niche. Most resources do not come nicely packaged in species-sized bits, but in amorphous glumps that can be divided in a multitude of ways. A highly generalized opportunist by definition can acquire multiple "glumps," but with time, more specialized species may evolve the necessarily adaptations to capture these packages. More importantly, however, niches are created by the organisms that occupy them through adaptations to the physical environment and to the other species with which they interact. A bird nest, a termite mound, or a clam burrow all represent ways in which individuals actively modify their environments. They are examples of species actively constructing the niche they inhabit. These constructions do far more however. An ant colony or a prairie-dog town also creates niches for other species. Aphids taken as slaves, predatory wasps, and other animals all find homes within ant colonies, as do beetles, snakes, and ferrets with prairie dogs. The construction of niches for other species is the most interesting part of the process, for this positive feedback allows species diversity to grow. Ecologists are not unaware of this process. Forest succession—from disturbed field, through shrubs, to pioneer trees, and eventually the "climax" forest of mature, slow-growing trees—has often been viewed as a process where the early species facilitate the invasion of later members of the community. In this ecological example, the species already exist and are migrating into a new area.

This process of niche construction means that rather than refilling an ecospace of niches that exists independent of the species that occupy them, the process of recovery builds the niches. During a recovery after a mass extinction, the total number of possible species is not fixed as in the chessboard model. Positive feedback allows the total number of possible species to expand as diversity grows. (For ecologists, this translates into an expansion of the carrying capacity of the ecosystem.) In my view, possibly the least understood part of the interaction of ecology and evolution is this process of niche creation, both during adaptive radiations and in the aftermath of mass extinctions.[25]

Recovery had only begun by the close of the Early Triassic, and biodiversity continued to climb through the remainder of the Triassic. Ecosystems were radically different on both land and in the sea as the effects of the extinction persisted for hundreds of millions of years. These longer-term effects of the extinction are the topic of the final chapter.

CHAPTER 10

The Paradox of the Permo-Triassic

Miocidaris keyserlyi was the end of a long tradition of Paleozoic sea urchins stretching back to the Ordovician, and almost the last of a lineage (figure 10.1). Sea urchins nearly became as extinct as their Paleozoic cousins, the blastoids. But two species, *Miocidaris* and a close relative, survived the trials of the Permo-Triassic boundary. From these two pioneering species came the incredible diversity of all modern sea urchins: the round, delicately spined *Diadema* of western North America; the great slate-pencil urchin of Hawaii, *Heterocentrotus mammillatus*, with long orange spines as thick as a pencil, the flattened discs of sand dollars, and the bilaterally symmetrical biscuits of the heart urchins (figure 10.2). Echinoids are one of the most significant members of today's shallow seas. Some actively erode coral reefs as they feed, while the grazing behavior of others is an important check on the growth of algae. Yet echinoids came within a species or two of extinction. Why did they survive, and how did they come to be such vital players in modern oceans?

Miocidaris was a geographically widespread echinoid genus and may have been highly opportunistic, with rapid development and reproduction allowing it to quickly expand when opportunities

Figure 10.1 *Miocidaris keyserlyi*, one of the two surviving sea urchin lineages and the progenitor of the Mesozoic radiation of the group.

permitted. These characteristics probably aided its survival during the mass extinction. But other features of *Miocidaris* made it utterly unlike any of the other seven known genera of Permian echinoids. Starfish, brittle stars, crinoids, echinoids, and almost all other echinoderms have five rays on the surface of the test, or skeleton. This pentameral symmetry is formed by the ambulacral plates that cover the tube feet (and are thus equivalent to the underside of the arms of a starfish). On a sea urchin, interambulacral plates lie between the ambulacral plates. For most Paleozoic echinoids the number of columns of interambulacral plates generally varies from one to five. In *Miocidaris*, however, the number is fixed at two. All post-Paleozoic echinoids inherited this situation and are constrained to have only two files of interambulacral plates. Why something did not evolve is generally an unanswerable question, so we will never understand why some greater variety of numbers of ambulacral plates did not re-evolve in the post-Paleozoic. The enormous diversity of the modern echinoids suggests that this architectural constraint had little impact on their ecological or taxonomic success.

Now, I could argue that this arrangement of plates somehow contributed to the survival of *Miocidaris* and changed the course of history (at least echinoid history). Somehow this arrangement of plates might have unlocked some survival advantage absent

Figure 10.2 From a relatively inauspicious beginning in the Early Triassic, sea urchins have diversified into a surprising array of morphologies, as shown in this montage of living urchins. All species are from the Florida-Caribbean area. Photographs courtesy of David Pawson (NMNH).

from the other species. Since only two lineages survived, and they alone had this unique feature, surely it must account for their survival. Alternatively I could argue that the survival of *Miocidaris* was luck and nothing more. Some clades disappeared during the extinction, others survived, and *Miocidaris* had enough individuals in the right place at the right time. In dealing with only two species, chance alone may be all the explanation required, or even possible. Another option, in some sense intermediate between the two, is that *Miocidaris* survived because of some other adaptation such as its opportunistic habits, and that the particular architecture of the test was simply an indirect and fortuitous consequence. With only two species crossing the boundary, there is no way to rigorously test these alternatives. Forced to choose, I would guess that *Miocidaris* survived through pure, dumb luck. What does intrigue me is the possibility, perhaps even the probability, that the survival and eventual triumph of a major clade of marine organisms was due to nothing but the luck of the draw.

The proliferation of such a bewildering variety of echinoids from *Miocidaris* mirrors the situation in many other clades. During the Paleozoic probably fewer than one in ten gastropod species were predatory carnivores. The vast majority were slowly searching through the muck for organic debris, or were suspension feeders, much like the more common brachiopods and bryozoans. A few were probably grazers, cropping algae. Contrast that with the situation today where as many as 90% of all living gastropod species are predators, many using highly poisonous neurotoxins to subdue their prey, including the odd shell collector. Some groups of bivalves developed the ability to fuse the margins of the mantle, the fleshy covering on the inside of the shell. Mantle fusion allowed the formation of tubes, or siphons, allowing bivalves to escape the surface by burrowing deep within the sediment. Siphons were the tenuous lifelines to the surface through which clams could pump water in, filter it for food and oxygen, and expel it back to the surface. So many other groups underwent similar revolutions in their mode of life that John Phillips viewed the Mesozoic as a second creation of animal life (chapter 2). The removal of the dominant Paleozoic groups allowed many previously minor groups to flourish in new and unexpected ways.

In this final chapter I will explore the long-term implications of the mass extinction for the history of life. It may seem self-evident that a mass extinction as large as the end-Permian event must have changed the course of history. Indeed, the work of Phillips and Jack Sepkoski described in chapter 2 appears to be prima facie evidence for the pervasive influence of the mass extinction. Yet Sepkoski, one of the foremost students of such issues, long championed the opposite view: that those groups that dominate the oceans today were already expanding well before the end-Permian extinction and would have triumphed by now even without the extinction. In other words, the mass extinction may have accelerated the process, but it did little to influence the outcome. The alternative view is that the ecology of Mesozoic and Cenozoic communities largely reflects the winners and losers of the Permian extinctions and the nature of the postextinction recovery. This debate remains unresolved, and is perhaps unresolvable, but raises the issue of what difference, if any, the extinction made in the history of life.

The new world that began with the recovery from end-Permian mass extinction was utterly different from the world of the Paleozoic. The transition began with the recovery in the Early and Middle Triassic but this was only the beginning of what Geerat Vermeij calls the "Mesozoic Marine Revolution."[1] Through the Triassic and Jurassic a host of new predators appeared, initiating adaptive responses on the part of their prey. Vermeij is a paleontologist at the University of California at Davis, and one of the world's foremost authorities on gastropods and the things that eat them. In the 1970s he became interested in the spread of spiny gastropods, of snails with narrow, slitlike apertures and the disappearance of some shell forms that had persisted for hundreds of millions of years. Suspecting that these patterns reflected a growing threat from predators, Vermeij began to chronicle how shells changed shape over time. Predation can only be inferred from such architectural evolution, so he looked for more direct evidence. Vermeij recognized that since many snails preyed on other snails and on clams by drilling into their shells, he could track the increase in predation by counting the frequency of drill holes through time. Although drill holes are not unknown in the

Paleozoic, and some shell-crushing animals had evolved during the Paleozoic, both groups increase considerably in abundance in the Mesozoic. Vermeij's work on gastropods, crabs, rays, and many other shell-crushing predators reveals that all of these groups began diversifying from the mid- to Late Triassic into the Cretaceous. The proliferation of these predators tracks the architectural changes in snails and clams that provide increased protection against shell crushing.

Chris McRobert's study of Triassic bivalves modifies this scenario, however. Bivalves that lived buried below the sea floor should have been more immune to shell-crushing predators, and have had a much lower rate of extinction than exposed bivalves. While this seems to support Vermeij's idea, McRobert's catalog of potential mollusc-eating predators shows that most only appeared near the end of the Triassic or later in the Mesozoic. Shell-crushing fish and cephalopods are a constant throughout the period, while ichthyosaurs, the voracious, dolphinlike marine reptiles, first appeared in the Early Triassic. Signs of drilling from gastropods do turn up among Triassic bivalves but are fairly rare, suggesting that this was not a major evolutionary factor. Predation was unlikely to have been an important evolutionary force until the Jurassic or Cretaceous. Many of the surviving bivalves in the Early Triassic were cosmopolitan, epifaunal generalists, and their disappearance by the mid-Triassic, coupled with a more rapid expansion of infaunal bivalves may account for the changing fortunes of bivalves in the two habitats, without any need to invoke predation.[2]

· · ·

Jack Sepkoski's characterization of the three great Evolutionary Faunas was part of his effort to understand the dynamics controlling the diversity of life in the oceans throughout the Phanerozoic. Far from being a simple descriptive enumeration of familial ranges, Jack used the data to develop several models.[3] These models used the same basic logistic equations discussed earlier, in chapter 9, but involved a separate equation to describe each of the Cambrian, Paleozoic, and Modern Evolutionary Faunas with different rates of increase and a maximum diversity for each. Both the rates of diversification and the maximum diversity were tuned to mimic the family diversity data. With the addition of perturbations to the model to mimic the effects of the great mass extinctions, the results were remarkably similar to the original data (figure 10.3). Some paleontologists criticized this work because it implicitly assumes that each evolutionary fauna behaved as some sort of evolutionary entity, although Jack pointed out that the faunas were simply statistical groupings of clades that displayed a similar diversity history through the Phanerozoic.

Look carefully at the behavior of the Paleozoic and Modern faunas through the Paleozoic in figure 10.3. The Modern Evolutionary Fauna appears to be expanding through the Paleozoic at the expense of the Paleozoic fauna. This trend is particularly prominent after the end-Devonian mass extinction in the mid-Paleozoic. The implication is that the clams, snails, and fish of the Modern fauna were already expanding their importance in the oceans, well before the end-Permian mass extinction, and would have come to dominate modern oceans *even in the absence of the end-Permian mass extinction.* In other words, this analysis suggests that the end-Permian mass extinction may have made little long-term difference to the history of life. Brachiopods and crinoids were losing, and they were losing long before the extinction. While the end-Permian extinction may have speeded things up a bit, it made no real difference to the final outcome.

The significance of the great mass extinctions in structuring Phanerozoic marine diversity was emphasized by an analysis of diversity from a functional and physiological perspective carried out by Richard Bambach and Andy Knoll in cooperation with Jack Sepkoski before his death. Using Jack's compendium of the stratigraphic ranges of 40,859 marine genera, they extended the earlier study of the role of physiology in Permo-Triassic extinction patterns. In this study the identified diversity patterns of two different functional subgroups (passive vs. active) and three different physiological groups (physiologically unbuffered, buffered, and indeterminate). The results show that the change in the proportion of passive vs. active animals changed at the end-Ordovician, end-Permian, and end-Cretaceous mass extinctions, with active groups

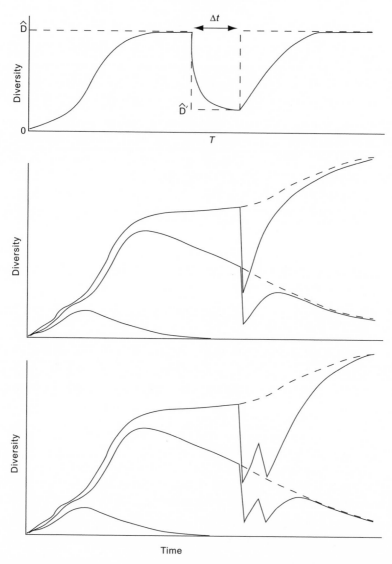

Figure 10.3 Jack Sepkoski's logistic model of Phanerozoic marine diversity, with diversity along the Y axis and time along the X axis. Top: a simple logistic curve with a single perturbation (lasting Δt time) followed by a logistic recovery. Middle: three coupled logistic equations mimic the diversity patterns of the Cambrian (1), Paleozoic (2), and Modern (3) Evolutionary Faunas with a single perturbation, corresponding to the end Permian mass extinction. Notice that the Paleozoic fauna is declining before the perturbation while the Modern fauna is already expanding. Thus the extinction has little impact on the final outcome. Bottom: a second pertubation is added. After Sepkoski (1984).

becoming progressively more dominant since the Silurian. For the physiologically buffered versus unbuffered groups, there is an even more striking shift at the end-Permian mass extinction, where the roughly 70% dominance of unbuffered taxa drops to about 45%. But the critical point is that the mass extinctions serve to divide long periods of stability. They did not impose a transient selective bias but fundamentally reordered the ecological structure of marine communities.[4]

To better understand what dynamics were driving this process, Jack and a former student, Arnie Miller, now a well-known paleontologist at the University of Cincinnati, compiled information on over five hundred Paleozoic marine communities from North America. They graded these localities according to their position across the shelf from near-shore environments all the way out to the outer continental slope, and according to the dominant groups within each community. Both the Paleozoic and Modern faunas arise during the Ordovician radiation, but the molluscan-dominated Modern fauna is in more near-shore habitats. Through the Paleozoic the Modern fauna progressively expands across the shelf, evidently driving the Paleozoic fauna into more offshore environments.[5] To my mind the real problem in interpreting the results comes during the Late Devonian extinctions. As I see the data, there seems to be an abrupt shift as a result of the extinction, with molluscan-dominated communities quickly capturing more territory immediately after the extinction. Jack and Arnie, however, interpreted the results as showing a progressive expansion of the Modern faunas.

One view of the onshore-offshore data supports Jack Sepkoski's view that the mass extinction made little difference in the eventual outcome of the battle between the Paleozoic and Modern Evolutionary Faunas. But the same data suggest the alternative possibility that the Late Devonian extinctions were the major factor in the late Paleozoic expansion of the Modern fauna. Although the analysis of functional groups and physiological characteristics described earlier in this chapter found no evidence of a significant change in the diversity structure associated with the Late Devonian extinctions, several other lines of evidence do suggest a shift in ecological patterns during the Devonian. Several paleontologists have described a "mid-Paleozoic precursor" to the Mesozoic Marine Revolution with an increase in evidence of predation and increased armoring and other protective structures for prey species.[6]

. . .

One of my favorite Sherlock Holmes stories is "Silver Blaze," in which Holmes draws the attention of Inspector Gregory to the curious incident of the dog in the nighttime. When Gregory observes that the dog did nothing in the nighttime, Holmes responds: "That was the curious incident." The curious incident in the Triassic was the lack of new phyla or classes. Although many new orders appeared, particularly among echinoids, all of the new architectures fit comfortably within existing groups; none are so unique, or so distinctive that they have been recognized as new phyla or classes. The elimination of over 90% of all marine species returned the world to a setting as barren as the late precambrian as illustrated by the profusion of stromatolites and other microbial fabrics. Yet while the late Neoproterozoic-Cambrian radiation of complex animals included the proliferation of several dozen new animal architectures (ranked as phyla) and many new classes, there was no such burst of morphological innovation in the Triassic. The lack of such innovation is particularly puzzling since most models of innovation invoke "empty ecospace" as a trigger for such innovation.

In one of my earliest papers, Jim Valentine, Jack Sepkoski, and I compiled the patterns of origination of phyla, classes, and orders and then compared these with the patterns of origination of short-lived families during the lower Paleozoic and Mesozoic[7] (at the time, the duration of the two intervals was thought to be essentially equivalent) (figure 10.4). We expected that the processes of morphological innovation that produce the wholly new architectures recognized as new phyla, classes, or orders should also be reflected at the family level, and that this should generate many short-lived families that should be concentrated during the intervals that produced new higher taxa. Figure 10.4 clearly shows the overwhelming predominance of new phyla and classes in the lowest Paleozoic.

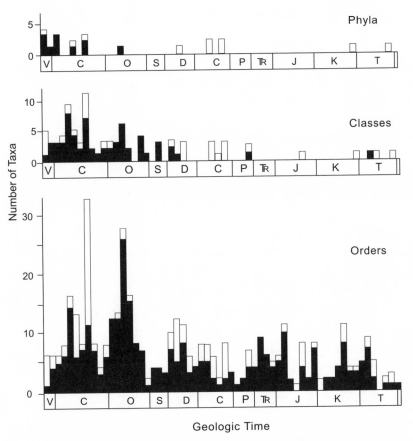

Figure 10.4 Most phyla originated in the late Neoproterozoic and Cambrian, classes of marine invertebrates during the lower Paleozoic, but orders originated throughout the Phanerozoic. There is some increase in the Triassic, but not a particularly noticeable one. Black boxes denote taxa that have durable skeletons and thus their first appearance in the fossil record is probably close to their time of origin; open boxes are taxa that are poorly skeletonized. Figure from Erwin et al. (1987).

Our hypothesis was an utter failure. There was no preferential concentration of short-lived families during the intervals in which new higher taxa appeared. Rather, the short-lived families were scattered throughout the two intervals. This was actually a far more interesting result than if we had been correct, for the pattern seems to suggest that the processes that facilitated the origination of radically novel morphologies was decoupled from events at the level of new families.

Many biologists object to this approach on the grounds that phyla, classes, and orders are not real entities. We were well aware of that criticism and were simply employing taxonomic ranks as proxies (albeit fairly imperfect ones) for morphological distinctiveness. Subsequently some more mathematically astute paleontologists have developed sophisticated means of directly measuring the degree of morphological change within clades.

The apparent absence of substantial morphological innovations following the end-Permian mass extinction has been explained in one of two ways. One possibility is that the breadth of the Cambrian radiation reflected the acquisition of new developmental capabilities. For example, the formation of appendages, segments, or eyes reflects the ability to coordinate the growth of highly specialized cell types into a complex morphological feature. To their considerable surprise, developmental biologists have recently discovered that animals as dissimilar and evolutionarily distant as flies and mice share a remarkable number of genes that control this developmental patterning. This can only mean that the last common ancestor of these two groups (as well as all other bilaterally symmetrical animals, from earthworms and molluscs to humans) possessed these same genes. Quite what this discovery implies for the Cambrian radiation is a highly contentious issue among developmental biologists and paleontologists. The implications for the Triassic are more obvious, however. If the radical reorganizations of morphology characterized as phyla and classes reflect reordering of developmental processes, and if (a very important if, incidentally) these developmental processes somehow became more rigid and less permissive of reorganization over time, then the magnitude of possible morphological innovation may have been constrained by a relatively inflexible developmental framework. This hypothesis requires that, as some paleontologists have held, major morphological innovations have become increasingly rare with time, since the Cambrian. While from a certain perspective it may be true that the diversity of vertebrates from fish to man is but rococo ornamentation on the brain and vertebral column that appeared in the Early Cambrian, it seems hard to deny the significance of bony fish, turtles, and placental mammals. Space does not permit delving into the arguments pro and con, but two facts seem clear. First, while there is a burst of innovation in the Cambrian, some new innovations have appeared through the Phanerozoic. Second, no one has been able to demonstrate any systematic decline in the production of such innovations over time, although the likelihood of their success may well have changed.

Ecological opportunity and the absence of competition are often invoked in explanations of evolutionary innovation. Darwin's finches on the Galapagos Islands are the paradigmatic example. Today the fourteen finches have a wide variety of beak types, from massive beaks well designed for crushing large fruit seeds to smaller, thinner beaks ideal for picking up tiny seeds. Since these finches are all descended from a single ancestral form blown in from Ecuador, this diversity of beak types is explained as specialization within an essentially open ecosystem. The abundance of different food types and the absence of any competitors have allowed different populations of finches to emphasize dissimilar food sources and evolve beaks specialized for each type of food. The many islands of the archipelago vary in their isolation from one another and in their abundance of different plants, two factors that aided the generation of so many different species.

Both the late Neoproterozoic and the Early Triassic can be viewed as the Galapagos Islands before the appearance of the first finch. In the ecological hypothesis, this lack of competition should generate many new species and clades. With sufficient ecological opportunity and time, the aftermath of a mass extinction should lead to widespread novelty, perhaps even rivaling the Cambrian radiation in breadth. Placental mammals exploded in basic architectures in the wake of the Cretaceous-Tertiary mass extinction. Despite some hints of placental mammals in the Late Cretaceous, the breadth of morphological innovation from bats to whales occurred in the 10 million years after the mass extinction. Given the greater magnitude of the end-Permian mass extinction, even more innovation might have been expected in the Early Triassic. The absence of extensive innovation could mean that erasing over 90% of all species was insufficient to allow new morphologies to evolve, or that such evolution requires more time than was available. Let us examine each alternative.

The 280 million years between the Cambrian and Permo-Triassic generated an evolutionary legacy for Early Triassic survivors utterly unlike the prior history of Cambrian animals. The Cambrian radiation began with the diversification of morphologically simple groups. Snails, clams, ammonoids, many arthropods, echinoids, bryozoans, even crinoids and some brachiopods were all present in the Early Triassic. True, the Paleozoic corals (the rugosans and tabulates) disappeared, but some unskeletonized sea anemones survived to produce the first scleractinian corals by the end of the Early Triassic. Trilobites finally became extinct, but the few Permian species were ecologically insignificant, and many other arthropod groups persisted.

This incredible variety of ways of making a living in the oceans can be divided into twenty different ecological strategies,[8] each distinguished by whether the animals are herbivores, carnivores, or suspension feeders; whether they swim, live on top of the sediment, or are buried within the sediment; and whether they are mobile or attached. Through the Phanerozoic, the number of strategies occupied increased, and through time, each class tends to diversify into a greater number of strategies. For example, whether true bivalves are present in the Cambrian is unclear, and they are not assigned to any of the twenty strategies for the Cambrian. By the middle and upper Paleozoic, bivalves have diversified into five different strategies, and ten by the post-Paleozoic. Only nine of the twenty strategies are inhabited in the Cambrian, and generally only a single class occupies each strategy. By the close of the Permian, some groups had acquired the adaptations to fill at least fourteen of the twenty possible strategies, and with the exception of the still sparsely inhabited burrowing environment, most strategies were practiced by three or more classes. Ecological opportunity sufficient to allow innovation would require evacuation of at least some of these strategies. Such was the distribution of survivors that well-adapted groups with diverse ecologies persisted in each strategy, allowing these incumbents to capitalize on new opportunities. Thus it seems likely that there simply were not enough unexploited ecological opportunities to catalyze the formation of novel morphologies.

Time may also have played a significant role in retarding innovation. The barren ecosystems of the earliest Triassic are certainly the closest analog to the late Neoproterozoic, with many strategies vacant or sparsely occupied. But this opportunity only lasted perhaps 4 million years, during the survival and recovery intervals. Innovation should be particularly prevalent during this interval. Scleractinian corals and other new orders appear, but the number is relatively few. The Cambrian radiation required far more than 4 million years, and the increasing diversity of Lazarus taxa and the proliferation of new species may have suppressed innovation before it proceeded as far as in the Cambrian. Of course the continuing environmental stress through the Early Triassic, as discussed in chapter 9, may also have stifled innovation.

Michael Foote at the University of Chicago compared the diversification of Paleozoic and post-Paleozoic crinoids, measuring sixty-nine characters on hundreds of different species from the Ordovician through the Tertiary.[9] After the end-Permian mass extinction, the number of crinoid species, genera, and families all peak in the Late Jurassic, but morphological breadth peaked about 50 million years earlier, in the Late Triassic. Yet this morphological peak occurred when species and generic diversity was only half that of the Late Jurassic. The Mesozoic rates of morphological diversification were very similar to those in the early Paleozoic, suggesting that at the levels of species, genera, and families, competition in post-Paleozoic communities was not sufficient to retard the production of new morphologies. Recall that Valentine, Sepkoski, and I found evidence of a discontinuity between the origination of new families and the higher taxonomic levels of order, so Foote's results are not necessarily incongruent. The absence of evidence for competition at lower levels does not preclude competition through ecological incumbency retarding major morphological innovations.

· · ·

For his presidential address to the Paleontological Society in 1996, Jack Sepkoski compared the current loss of species to past mass extinctions. Since we do not know how many species are actually alive today, and the estimates of how rapidly species are disappearing are even less precise, this was a chancy exercise. But by using the range of estimates, it is possible to gauge how long we can continue on our present course before we are well into a mass extinction. Many conservationists have been describing the current loss of biodiversity as "the sixth extinction" but Sepkoski concluded that it would be another 280 to 10,000 years before we reached levels of species loss comparable to the end-Permian mass extinction. This of course is based on comparison to marine extinction rates. On land our predecessors began decimating large mammals about 50,000 years ago in some parts of the world, and our current rapacious ways are best seen as a continuation of this trend.[10]

The purpose of such comparisons is to disquiet, anger, and motivate right-minded citizens to oppose the ungodly purveyors of death and destruction. This is all well and good, but it does help if you have the numbers right. As Jack Sepkoski well understood (but many conservation biologists do not) any comparison of fossil extinction rates to current estimates is inherently flawed because the data are so different. With certain obvious exceptions (passenger pigeons, mastodons, and saber-toothed tigers leap to mind), most of the species that humans have so thoughtlessly eliminated are local, often rare, and unlikely to be preserved in the fossil record. The late Allen Gentry, an indefatigable botanist, often illustrated his talks with a discussion of the loss of many plant species from an Ecuadorian cloud forest in the ten years between his visits. Accepting that this was not simply a sampling problem (in fact the most plausible explanation), the likelihood that any of these plants would have been preserved as fossils is effectively zero. The fossil record is overwhelmingly composed of common, geographically widespread, and easily preserved organisms, the forms most likely to win the incredible lottery of death, decay, and destruction they must pass through to achieve paleontological immortality. The past 600 million years of history could include dozens of mass extinctions of poorly preserved organisms like those described by Gentry and we would be none the wiser. It is far more appropriate to compare past mass extinctions to the number of species that have disappeared among common, widespread, and durable species. There is no political motivation for such a comparison, of course, as it would significantly lower the apparent similarity between past mass extinctions and the current situation.

Human impact on the environment is hardly limited to the observed loss of species. Population growth and the attendant capture of terrestrial and marine productivity, degradation of habitat and alteration of geochemical cycles of carbon, nitrogen, and phosphorus have also altered the number of species the earth can support. It seems inescapable that many species currently alive are what Dave Jablonski terms "Dead Clade Walking": many individuals are still alive but the species or clade is already doomed to extinction. Friends of mine have put enormous effort into the protection and captive breeding of the magnificent California condor, but with so few birds remaining, the likelihood that condors will by flying around the hills of central California in 1,000 years seems remote.

The inappropriateness of current events to past biodiversity crises is, paradoxically, the best reason for hope. Mass extinctions represent the destruction of the ecological fabric that permits communities to function. If we are indeed in the midst of a mass extinction, it may already be too late to do much to arrest this destruction, a destruction that will arguably include the extinction of our own species. And the lesson from biotic recoveries is that they take far longer than the recorded history of *Homo sapiens*, and occasionally far longer than the entire history of hominids. So the best hope and argument for those wanting to preserve some of the incredible diversity of life that surrounds us is that we are not yet in a mass extinction.

Notes

CHAPTER 1
INTRODUCTION

1. There is a long and fascinating history of geology and paleontology of the Guadalupe Mountains and the Permian reef complex, and this has been described in many places. Among the more useful publications on the reef are Wood et al. (1994), and of historical interest is the classic book by Norman Newell et al. (1953), and a historical overview of work on the reef by Rigby and Millward (1988). Wood's (1999) recent book on the evolution of reefs includes a discussion of the Permian reef complex and places it in the context of other reefs through geologic history.

CHAPTER 2
A CACOPHONY OF CAUSES

1. Only the briefest overview of these other mass extinctions is provided here, but more detailed discussions of mass extinctions can be found in Erwin (2001), Raup (1991), and a comprehensive discussion in Wignall and Hallam (1997). McGhee et al. (2004) discuss the evolutionary and ecological implications of the various mass extinctions. For specific events, see also Tanner et al. (2003) for a slightly different view of the end-Triassic mass extinction.

2. The original Alvarez hypothesis was published in Alvarez et al. (1980). Countless accounts of this research have been published in the past decade.

3. Sepkoski, 1979, 1981, 1984.

4. Bambach and Knoll, 2001.

5. Raup, 1991; Raup and Sepkoski, 1984, 1986.

6. Xu et al., 1985; Clark et al., 1986; Asaro et al., 1982; Kakuwa and Toyoda, 1996.

7. Orth and Quintana, 1991. The newer analysis is by Koeberl et al. (2004).

8. Mory et al., 2000.

9. Comment on ibid., by Reimold and Koeberl (2001) and response by Mory et al. (2001).

10. Becker et al., 2004.

11. Retallack et al., 1998.

12. See Becker et al. (2001). Poreda and Becker (2003) later reported the isolation of fullerenes with a similarly enriched helium gas component from a Permo-Triassic boundary section at Graphite Peak, Antarctica. Buseck (2002) provides a useful, general introduction to fullerenes.

13. Abbas et al., 2000.

14. Courtillot and Renne, 2003. Estimates of the size of the Siberian traps are also in Nikishin et al. (2002).

15. Valentine and Moores, 1970, 1973.

16. MacArthur and Wilson, 1967.

17. Schopf (1974) discussed the impact of sea-level change on biotic diversity in light of the species-area effect, while Simberloff (1974) described the linear nature of the relationship and showed that the change in diversity across the Permo-Triassic boundary could be explained by the then-current estimates of changes in sea level.

18. Stanley, 1988a, b.

19. The series of papers on anoxia began with Wignall and Hallam (1992, 1993) and then continued with Paul Wignall's research group (Wignall et al., 1995, 1998, 2005; Wignall and Twitchett, 1996; Twitchett et al., 2001).

20. Isozaki's work on deep-sea anoxia is described in Isozaki et al. (1995, 1997) and Musashi et al.(2001).

21. Grotzinger and Knoll, 1995; and Knoll et al., 1996.

CHAPTER 3
SOUTH CHINA INTERLUDE

1. Here is a bit of technical trivia that will bore most. The actual Permo-Triassic boundary does not coincide with the mass extinction, but lies about 12 cm above it in bed 27c. Geologists do not define the ends of intervals but rather their beginnings, and the formal definition of the beginning of the Triassic is the first occurrence of the conodont *Hindeodus parvus* at section D at Meishan. So if one were to be really pedantic this would be known as the Late Permian mass extinc-

tion rather than the Permo-Triassic mass extinction. That, however, is too pedantic even for (most) geologists.

2. The confidence-interval analysis was published by Jin et al. (2000b), and the radiometric ages by Bowring et al. (1998); the latter are discussed further in chapter 4. Charles Marshall's confidence-interval statistics were discussed by Marshall (1994), and Marshall and Ward (1996).

3. Holser and Magaritz, 1987. Holser and Magaritz were following a tradition in Permo-Triassic studies that extends back to Tom Schopf's work in the early 1970s; for more detail, please see chapter 5 of Erwin (1993).

4. See chapter 2, note 19.

5. Chen et al., 1998; and Tong et al., 1999.

6. Steve Kershaw from Brunel University in England is the leading expert on these anomolous crusts, and Kershaw et al. (1999, 2002) are the critical papers. Dan Lehrman has found similar carbonates in his work on the Grand Bank of Guizhou (Lehrman, 1999; Lehrmann et al., 2001), considerably extending the geographic range of these fabrics within south China. These crusts have been studied for years and have variously been ascribed to evaporites, calcretes, or other causes, but these claims are not supported by more recent studies.

7. See Heydari et al., 2003. The sediments also indicate that the latest Permian rocks were deposited in deep, oxygenated waters (no evidence of deep-water anoxia or sluggish oceanic circulation), but Heydari and colleagues argue that these unusual carbonates appear to have been deposited in shallow water, revealing a rapid, significant drop in sea level at the boundary. In this argument they rely on outdated views of sea-level change across the Permo-Triassic boundary, and it is unclear if such a major change in sea level is actually required. Furthermore, earliest Triassic sediments were clearly deposited in deep water. The Japanese occurrences were reported by Sano and Nakashima (1997).

8. Aldridge et al., 1993.

9. There is a large literature on conodont biostratigraphy of the Permo-Triassic boundary, with some of the significant papers including Kozur, 1996; Orchard, 1998; and Sweet, 1992.

CHAPTER 4
IT'S A MATTER OF TIME

1. Bowring et al., 1998. Further discussion of the Meishan geochronology may be found in Jin et al. (2000b) and Erwin et al. (2002).

2. Mundil's results are discussed in Metcalfe et al. (2001) and Mundil et al. (2001). Comparison of the Bowring et al. (1998) and Mundil et al. (2001) results can be found in Erwin et al. (2002).

3. Mundil et al., 2004.

4. Rampino et al., 2000.

5. The major papers on dating the Siberian volcanics are Renne and Basu (1992); Campbell (1992); Renne et al. (1995); Basu et al. (1995); and Kamo et al. (1996, 2000). Reichow et al. (2002) provide new argon-argon dates on the west Siberian Basin extending the flood basalts far to the west and doubling their extent.

6. Kamo et al., 1996, 2000, 2003.

7. Lind et al., 1994.

8. Zhou et al., 2002. Lo et al. (2002) dated the Emeishan volcanics via argon-argon dating, producing estimates of about 251–253 million years ago. These dates appear to overlap with the PT boundary, but there are well-known Late Permian fossils lying on top of the Emeishan volcanics, which is a bit of a problem for Lo et al. Argon-argon dates also do not correspond precisely with uranium-lead dates, and there may be a systematic error producing the much younger argon dates relative to the uranium-lead dates of Zhou et al. That the Emeishan flood basalt was produced by a mantle plume has been shown by extensive domal uplift prior to the eruption: He et al. (2003); see also discussion by Xu et al. (2004) and Ali et al. (2005).

9. Szurlies et al. (2003) provide a recent review of the magnetostratigraphy of the boundary interval, with emphasis on the Early Triassic. Jin et al. (2000a) is an earlier but still very useful review, integrated with recent biostratigaphy. Haag and Heller (1991) remains the classic reference on the subject.

CHAPTER 5
FILTER FEEDING FAILS

1. Sepkoski, 1984.

2. Erwin, 1993. This volume is now dated in many areas, for which I am grateful. The obsolescence of the book reveals how rapidly our knowledge of the Permo-Triassic events has improved.

3. Stanley and Yang, 1994; and Jin et al., 1994. The Guadalupian brachiopod extinction was analyzed by Shen and Shi (2002).

4. Signor and Lipps, 1982. Until the Alvarez-impact hypothesis, paleontologists had not really thought through the implications of a catastrophic event.

5. See for example the discussion by Shi and Shen (2000) on the extinction patterns among Permian brachiopods and the likely involvement of a drop in sea level.

6. Villier and Korn, 2004.

7. The work of Guirong Shi, Neil Archbold, and their students is described in Shen et al. (2000), and Shen and Shi (2002).

8. Sandy Carlson's studies of brachiopod phylogeny and its implications for the mass extinction are discussed in Carlson (1991).

9. Wang and Sugiyama (2000) discuss global coral diversity, and Fedorowski (1989) studied the biogeographic distribution of Permian corals. Late Permian

corals in Spitsbergen and Greenland are discussed by Ezaki et al. (1994) and Ezaki (1997a, b).

10. Zhou, Z. et al., 1999; and Villier and Korn, 2004.

11. Erwin, 1996, and unpublished data.

12. Weidlich (2002) provides a recent summary of the structure of Permian reefs. Flügel and Kiessling (2002) discuss the impact of mass extinctions, including the two Permian events, on reefs and on carbonate production. Flügel and Reinhardt (1989) compared some latest Permian reefs in China and Greece, while Shen et al. (1998) describe one of the youngest Permian reefs from Hunan Province, China. Wood (1999) discusses the importance of decoupling the formation of reefs and the existence of carbonate platforms from the existence of reef-building organisms. The one good comparison of reef and nonreef extinction patterns is Raup and Boyajian (1988), and more recently the study cited above by Flügel and Kiessling.

13. Although Richard Bambach promises me the "long version," full of new analyses and additional documentation, to date the only published version of this hypothesis is Knoll et al. (1996). Further discussion of the implications of the idea for Phanerozoic diversity can be found in Bambach et al. (2002).

14. The analysis of the Meishan data was reported in Jin et al. (2000b), with Wang Yue from Jin's lab doing most of the collection and analysis of the data— a prodigious amount of work. The study covers the records of foraminifera, radiolarians, rugose corals, bryozoans, brachiopods, bivalves, cephalopods, gastropods, ostracods, trilobites, conodonts, fish, calcareous algae, and miscellaneous other forms. Charles Marshall's approach is discussed in Marshall (1994) and Marshall and Ward (1996).

15. Rampino and Adler, 1998.

16. Raup, 1991.

CHAPTER 6
SOUTH AFRICAN EDEN

1. Palmer, 1966.

2. Late Permian vertebrate ecosystems are discussed in Behrensmeyer et al. (1992), and the evolutionary relationships by Rubidge and Sidor (2001).

3. The new study of the phylogenetic relationships is by Modesto et al. (2001), and further evidence is presented by Modesto et al. (2003).

4. The collaborative study by Bruce and his colleagues was published as Rubidge (1995). The transition from *Dicynodonn* to *Lystrosaurus* has long formed the boundary between the Permian and Triassic in the Karoo but also in many other parts of the world. Ken Angielczyk, a young vertebrate paleobiologist at the Museum of Paleontology at Berkeley and Andrey Kurkin from the Palaeontological Institute in Moscow performed a phylogenetic analysis of some South African and Russian *Dicynodon*s. "*Dicynodon*" turns out to have evolved several

times in their analysis, and is not a monophyletic group. Although the genus has been recorded from Russia, their results suggest that *Dicynodon* did not occur in Russia, and so of course it cannot be used for correlations between Russia and South Africa. See Angielczyk and Kurkin (2003).

5. James Kitching died on 24 December 2003, the last of the giants of Karoo paleontology. For details of Kitch's life, I am indebted to Bruce Rubidge for copies of the the the obituaries he wrote with Michael A. Raath.

6. MacLeod et al., 2000; see also Ward et al., 2005.

7. Benton et al., 2004. The paper by Benton and colleagues also details the ecological changes associated with the extinction. They also carefully analyzed the effects of changes in sampling effort and sample size, and found no evidence that such artifacts produced the apparent extinction. Details of the biotic recovery are discussed in chapter 9. Tverdokhlebov et al. (2005) provides more details on the vertebrates and their sedimentology.

8. Smith and Ward, 2001; and Ward et al., 2000. Retallack et al. (2003) discusses the fossil soils of the Karoo and the various environments in which they formed. They argue for a distinctive change in the character of fossil soils across the Permo-Triassic boundary, suggestive of a change from an arid environment with highly seasonal rainfall to a warmer, more semi-arid and wetter climate. They emphasize, however, that this climatic shift seems insufficient to induce an extinction of perhaps 88% of fossil vertebrate genera, and suggest that massive release of methane produced an atmosphere with too much carbon dioxide and too little oxygen. Retallack et al. (2003) was criticized by Engoren (2004) based on the fact that the proposed drop in the volume of oxygen from 30% to 12% was similar to the altitudinal difference to which animals can readily adapt today. Engoren also pointed out that the relevant variable is the partial pressure of oxygen, not simply the volume percent. See also the response by Retallack (2004).

9. The work of Michael Benton and his colleagues in Russia is discussed in Newell et al. (1999) and Tverdokhlebov et al. (2005).

10. Sarkar et al., 2003.

11. Michaelsen, 2002.

12. Retallack et al., 1996.

13. Retallack, 1995.

14. Knoll, 1984; Rees, 2002.

15. McLoughin et al., 1997.

16. Conrad Labandeira recently published an overview of insect extinctions (Labandeira in press). Labandeira and Sepkoski (1993) remains an excellent summary of the impact of the extinction on insects.

17. Eshet et al., 1995; and Visscher et al., 1996.

18. Steiner et al., 2003. The section studied at Carelton Heights is lithologically similar to PT boundary sections at Bethulie and Lootsperg Pass, but has not been well studied. The fungal spike is found between the last *Dicynodon* and the first *Lystrosaurus*, and coincident with the last typical Permian gymnosperm pollen. Steiner et al. claim on the basis of sedimentation rates that the extinction

occurred in less than 40,000 years, but there are too few age constraints to have any confidence in this claim.

19. Afonin, 2001.

20. Foster et al., 2002. The *Reduviasporonites* had −13C ratio of about −32 per mil, far to light for it to have been feeding off woody vegetation at −22 per mil.

21. Twitchett et al., 2001; and Looy et al., 2001.

CHAPTER 7

THE PERILS OF PERMIAN SEAS

1. The carbon isotopic shift at Meishan is discussed in Jin et al. (2000b), and in more detail by Cao et al. (2002); see also Payne et al. (2004).

2. The results of the Gartnerkofel study were first presented in a paper in *Nature* (Holser et al., 1989) followed by a detailed monograph (Holser et al., 1991).

3. Baud et al., 1989.

4. The shift in the organic carbon record is detailed in Wignall et al. (1998), Krull et al. (2000), Cao et al. (2002).

5. For the Alps, see Magaritz et al. (1992), for Meishan see Cao et al. (2002), and for India see Musashi et al. (2001). The suggestion of latitudinal variation in the magnitude of the carbon isotope shift was made by Krull et al. (2000), but they included some high northern latitude records that have been shown to be artifacts (see discussion in Erwin et al. [2002]). Sephton et al. (2002) reported on the analysis of plant cuticles and the simultaneous changes in both carbonate and organic carbon from the southern Alps.

6. These results are discussed in de Wit et al. (2002). Additional analyses of sections in India suggest that the duration of the multiple excursions may have been as long as 500,000 years (Sarkar et al., 2003).

7. Foster et al. (1998) discusses the problems of fluctuating sources of carbon.

8. The Spitzy and Degans equations and the early methane story are detailed in Erwin (1993). The early estimates of methane volume are in Dickens et al. (1997, 2000); the more recent results at Hydrate Ridge were reported by Milkov et al. (2003). Much more sophisticated models of the changes to carbon isotopes have been published in the past few years, including Berner (2002). Buffet and Archer (2005) described a model of changes in the global methane inventory as a result of changing primary productivity, oceanic warming, and changing amounts of oxygen in the water. Each of these may have changed in the latest Permian but further modeling is required to see how, and whether, such changes could have influenced the mass extinction.

9. Svensen et al., 2004. This very intriguing hypothesis has yet to receive detailed scrutiny, but it does suggest means to produce methane and achieve the magnitude of the change in carbon isotopes that is more directly linked to the Siberian volcanic eruptions.

10. Sephton et al. (2002) suggested that soil carbon might explain an isotopic shift after destruction of plants increased erosion.

11. Sulfur isotopes across the Permo-Triassic boundary are discussed by Kajiwara et al. (1994) for the Japanese Sasayama locality. Changes in the terrestrial sulfur patterns in the Karoo of South Africa are chronicled by Maruoka et al. (2003), who relied upon a constant ratio of organic carbon to sulfide through the sequence as evidence that sulfate must have been added. The effects of Mount Pinatubo were discussed by Newhall et al. (2002).

12. Sulfur isotopes across the Permo-Triassic boundary were discussed by Kaiho et al. (2001) for the Meishan locality. Kaiho and colleagues assumed that they could calculate the radius of the impact crater by assuming that the amount of sulfur was somewhere between that found in average mantle material, and a Hawaiian island basalt. Yet the data from Mount Pinatubo shows that for massive explosive volcanism, vastly greater amounts of sulfur can be released than a geologist would predict. Nonetheless, they produce an estimate of a crater 600–1,200 kilometers in diameter, and a 15–60-kilometer-impact object, with the smaller object being a high-velocity comet, and the larger part of the range of a meteorite.

13. The average mantle value of $^{86}Sr/\,^{87}Sr$ is 0.703, while the average value of fractionated granitic crust is 0.710, and can be even higher in an older continental crust. Kaiho et al. (2001) identified a drop in $^{86}Sr/\,^{87}Sr$ from 0.715 to 0.708 at the top of bed 24 at Meishan, then an increase to 0.733 by bed 27. The longer-term pattern was identified by Martin and Macdougall (1995).

14. Sediments document the shift to anoxic conditions, and these have been described from several Japanese sections by Kakuwa (1996) and from British Columbia and Japan by Isozaki (1997). The new results from Hovea-3 and Meishan were reported by Grice et al. (2005), and evidence of sulfidic deep water in Greenland by Nielsen and Shen (2004).

15. The Wignall and Hallam studies on anoxic conditions are detailed in Wignall and Hallam (1992, 1993) and Wignall et al. (1995). Advances in conodont biostratigraphy provide a finer scale division of time, allowing Wignall and Hallam to track the increased anoxia through conodont zones. Such detailed records revealed that anoxic sedimentation began below the *Hindeodus parvus* conodont zone of the earliest Triassic in south China, and slightly later in Kashmir. But in the Salt Range of Pakistan, sea level does not change until two conodont zones after *parvus*, when anoxia appears as well. The differing pattern of appearance of anoxia in different regions is documented in Wignall et al. (1996). Anoxia in Spitsbergen sediments was discussed by Wignall and Twitchett (1996).

16. Beauchamp's work has been presented in a series of papers, of which Beauchamp (1994), Beauchamp and Desrochers (1997), and Beauchamp and Baud (2002) are the most relevant.

17. Beauchamp and Baud, 2002.

18. Some of these other explanations of silica patterns were explored by Kidder and Erwin (2001) and Racki (1999).

CHAPTER 8
DENOUEMENT

1. Visscher et al., 2004.

2. Courtillot et al., 2003.

3. Papers by Fedorenko et al. (1997, 2000); Czmanske et al. (1998); and Zolotukin and Al'Mukamudi (1998).

4. Renne et al. (1995) explored the patterns of uplift. For Fedorenko et al., see note 3. Papers by the Reichow group questioning the conclusions of the Czemanske group include Reichow et al. (2005) and Saunders et al. (2005).

5. Elkins-Tanton and Hager, 2000.

6. The possible origin of the Siberian flood basalts by continental rifting has been discussed by Puffer (2001), Courtillot and Renne (2003), and Nikishin et al. (2002). Various papers have discussed the direct effects of volcanism, particularly the prospect of acid rain caused by massive amounts of sulfate aerosols. See Renne et al. (1995).

7. Wignall, 2001.

8. Brown, 2002.

9. Gene Shoemaker's analysis of the cratering record was published in Shoemaker (1983). See also Abbot and Isley (2002).

10. Abbot and Isley, 2002. Gilkson (2003) provides very cogent criticism on Aboot and Isley (2003); Abbott and Isley (2003) provide a response. The limited number of large craters is from Ivanov and Melosh (2003).

11. Ivanov and Melosh, 2003.

12. See note 5. Jones et al. (2002) also built simulations of the effect of impact on the formation of large igneous provinces, with conclusions similar to those of Elkins-Taunton and Hager. They conclude that impacts are a plausible source of large flood basalts of the magnitude of the Siberian traps. More interestingly they observe that such massive craters will "auto-obliterate," destroying themselves as material flows back into the crater. This may explain the lack of large craters on the earth in comparison to what might be expected.

13. Morgan et al., 2004.

14. Poreda and Becker, 2003.

15. Becker et al., 2004. Some of the response by other impact specialists is discussed in Kerr (2004). Andy Glikson provided me with an unpublished analysis of the Bedout material that he prepared and discussed his objections to the Bedout structure as an impact structure in several emails. The pointed criticisms of the Becker et al. paper were published in the 22 October 2004 *Science*, as Glikson (2004), Wignall et al. (2004), and Renne et al. (2004) with responses by Becker's group (Becker et al. [2004b, c, d]). Other evidence suggestive of impact comes from work by David Agresti and Tom Wdowiak at the University of Alabama at Birmingham, who compared the base of the ash at the Meishan boundary to other volcanic ash beds above the contact, and to the Cretaceous-Tertiary boundary. Wdowiak specializes on Mössbauer spectroscopy, a technique that ana-

lyzes the change in magnetic properties of a mineral as its temperature changes (a process with the euphonious name of superparamagnetism). Iron particles are very sensitive to this technique, and the Alabama group had previously shown that Cretaceous-Tertiary boundary layers contained exceedingly fine iron particles. Results from the Meishan boundary samples were remarkably similar to the Cretaceous-Tertiary boundary, suggesting the presence of an extremely fine iron-rich layer, possibly debris of an iron-rich meteorite. This iron-rich layer probably produced the reddish stain at the top of bed 24 at Meishan. Agresti and Wdowiak, NASA Ames Astrobiology meeting (2002).

16. Basu et al. (2003); see also the news article in the same issue by Kerr (2003). A 2004 abstract by Petaev et al. provides additional details on some iron-silicon-aluminum spheres from Antarctica. Some of these silica and aluminum glasses have captured magnetite or silica grains, evidently during very rapid cooling in a hot, gaseous cloud. The large amounts of aluminum and titanium and the absence of magnesium, calcium, sodium, or potassium strongly point to an extraterrestrial rather than a volcanic origin.

17. Bowring et al., 1998. We calculated that a 10-kilometer-diameter comet could contain sufficient methane at -85 $\delta^{13}C$ to produce the carbon shift.

18. Wignall and Hallam 1992, 1993; Wignall et al., 1995, 1998; Wignall and Twitchett, 1996; Twitchett et al., 2001. Additional support for widespread anoxia comes from Korte et al. (2004). See also Kidder and Worsley (2004).

19. Weidlich et al., 2003; Brookfield et al., 2003.

20. Kennedy et al., 2002.

21. The full model is presented in Knoll et al. (1996), and Grotzinger and Knoll (1995) discuss in greater detail the anomalous carbonates of the Late Permian and their possible formation as inorganic precipitates from seawater supersaturated in carbonate. Ryskin (2003) proposed a similar model but with considerable methane in the deep stagnant water mass, in addition to carbon dioxide and hydrogen sulfide. This gathered a brief burst of media attention, probably spurred by the cinematic possibilities. This idea suffers both from the various problems of the Knoll model as well as the new, lower estimates of methane production rates.

22. Hortinski et al. (2001) used a three-dimensional general circulation model to simulate Permian ocean circulation and linked this to a biogeochemical model of the behavior of phosphate and oxygen. The model developed at MIT by Zhange et al. (2001) used a different model and different parameters. Some of the differences in approach between the two models are discussed in a comment by Hotinski et al. (2002) and a response by Zhang et al. (2003). Kidder and Worsely (2004) present a different view of the climate dynamics and argue that the haline circulation model of Zhange et al. may have been more persistent.

23. Sheldon and Retallack, 2002.

24. Wignall, 2001.

25. Krull and Retallack, 2000; and Krull et al., 2000.

CHAPTER 9
RESURRECTION AND RECOVERY

1. See, for example, Chen et al., 2002.

2. Batten, 1973. Jablonski's appellation is in Jablonski (1986).

3. See discussion of Kalkowsky's stromatolites in Paul and Peryt (2000). As my colleague John Grotzinger from MIT has noted, although most paleontologists continue to assume all stromatolites are microbial structures, such structures can also be produced by inorganic precipitation if the chemistry of the waters is appropriate. This can make separating microbial and inorganic origins quite difficult. An unresolved question about the earliest Triassic stromatolites is how many are truly of biologic origin and how many represent unusual inorganic precipitates and thus reflects something about ocean chemistry of the time. Pruss and Bottjer (2004) documents the geographic extent of Early Triassic stromatolites. See also Pruss et al. (2004).

4. Pruss et al., 2004.

5. The diversity of the Early Triassic marine assemblages in the western United States was first studied by Schubert and Bottjer (1995) and more recently by Rodland and Bottjer (2001).

6. Twitchett and Wignall, 1996; Twitchett, 1999; Twitchett et al., 2004.

7. Krystyn et al., 2003.

8. Woods et al., 1999; Woods and Bottjer, 2000.

9. McGowan, 2004.

10. Fraser and Bottjer, 2004.

11. Payne, 2004.

12. The Guizhou bank is described by Lehrman, 1999. An excellent review of the patterns of Phanerozoic reef crises and recovery is Flügel and Kiessling (2002), which describes the earliest Triassic reefs. See also Pruss and Bottjer (2004).

13. Stanley (2003) discusses the evolutionary history of the scleractinian corals and their first appearance in the Triassic.

14. Flügel (2002) summarizes the evolution of Triassic reefs, and Senowbari-Daryan et al. (1993) report on the recovery of Middle Triassic reefs in the Alps of Italy.

15. Retallack, 1997.

16. Looy et al., 1999, 2001.

17. Wang, 1996.

18. Veevers et al., 1994; Retallack et al., 1996.

19. Retallack, 1999; and Retallack and Krull, 1999. The berthierine in Early Triassic soils is described by Sheldon and Retallack (2002).

20. Benton et al., 2004.

21. The carbon isotope record for the Early Triassic is detailed in Payne et al. (2004). A colleague of mine at Wuhan was conducting a similar analysis at the Chaohu localities, one of the best preserved Chinese Early Triassic sections: Tong

et al., Lower Triassic carbon isotope stratigraphy in Chaohu, Anhui: Implications for biotic and ecological recovery. Submitted.

22. Wignall and Benton (1999) criticized an earlier discussion of potential preservational problems. They counted the number of shallow marine formations and claimed that the lack of substantive change through the Permo-Triassic boundary allowed one to reject preservational problems. In fact, this data completely fails to address the issue since the relevant issue is the quality of preservation and the extent of silicification.

23. Solé et al., 2002. In the initial model no species are omnivores, carnivores, and herbivores, although Jose Montoya, a student of Ricard Solé's is developing a model to include this added complexity.

24. Chase and Leibold, 2003.

25. Laland et al., 1999; 2000; Odling-Smee et al., 2003.

CHAPTER 10
THE PARADOX OF THE PERMO-TRIASSIC

1. Vermeij, 1977, 1987; see also Roy (1994) for a discussion of a specific example from aporrhaid gastropods.

2. McRoberts, 2001.

3. Sepksoki, 1981, 1984.

4. Bambach et al., 2004.

5. Sepkoski and Miller, 1985. See also Sepkoski (1987, 1991).

6. The mid-Paleozoic precursor to the Mesozoic marine revolution is discussed by Signor and Brett (1984), Bambach (1999), and Baumiller and Gahn (2004).

7. Erwin et al., 1987.

8. Bambach, 1995.

9. Foote, 1995, 1996.

10. Sepkoski, 1997.

References

Abbas, S. A. Abbas, and S. Mohanty. 2000. "Anoxia during the Late Permian binary mass extinction and dark matter." *Current Science* 78:1290–92.

Abbot, D. H., and A. E. Isley. 2002. "Extraterrestrial influences on mantle plume activity. *Earth and Planetary Science Letters*" 205:53–62.

———. 2003. "Reply to Comment on 'Extraterrestrial influences on mantle plume activity' by Andrew Gilkson." *Earth and Planetary Science Letters* 215:429–32.

Afonin, S. A., S. S. Barinova, and V. A. Krassilov. 2001. "A bloom of *Tympanicysta* Balme (green algae of zygnematalean affinities) at the Permian-Triassic boundary." *Geodiversitas* 23:481–87.

Aldridge, R. J., D.E.G. Briggs, M. P. Smith, E.N.K. Clarkson, and N.D.L.Clark. 1993. "The anatomy of conodonts." *Philosophical Transactions of the Royal Soiety of London*, B 340:403–21.

Ali, J. R., G. M. Thompson, M.-F. Zhou, and X. Song. 2005. "Emeishan large igneous province, SW China." *Lithos* 79:475–89.

Alvarez, L. W., W. Alvarez, F. Asaro, and H. V. Michel. 1980. "Extraterrestrial cause for the Cretaceous-Tertiary extinction." *Science* 208:1095–1108.

Angielczyk, K. D., and A. A. Kurkin. 2003. "Has the utility of *Dicynodon* for Late Permian terrestrial biostratigraphy been overstated?" *Geology* 31:363–66.

Asaro, F., L. W. Alvarez, W. Alvarez, and H. V. Michel. 1982. "Geochemical anomalies near the Eocene/Oligocene and Permian/Triassic boundaries," pp. 517–28 in L. T. Silver and P. H. Schultz, eds. *Geological Implications of Impact Hypothesis of Large Asteroids and Comets on the Earth, Special Paper.* Geological Society of America, Boulder, CO.

Bains, S., R. M. Corfield, and R. D. Norris. 1999. "Mechanisms of climate warming at the end of the Paleocene." *Science* 285:724–27.

Bambach, R. K. 1985. "Classes and adaptive variety: The ecology of diversification in marine fauans through the Phanerozoic," pp. 191–53 in J. W. Valentine, ed. *Phanerozoic Diversity Patterns.* Princeton University Press, Princeton, NJ.

———. 1999. "Energetics in the global marine fauna: a connection between terrestrial diversification and change in the marine biosphere." *Geobios* 32:131–44.

Bambach, R. K., A. H. Knoll, and J. J. Sepkoski, Jr. 2002. "Anatomical and ecological constraints on Phanerozoic animal diversity in the marine realm." Proceedings of the National Academy of Sciences, USA 99:6854–59.

Basu, A. R., R. J. Poreda, P. R. Renne, F. Teichmann, Y. R. Vasiliev, N. V. Sobolev, and B. D. Turrin. 1995. "High-³He plume origin and temporal-spatial evolution of the Siberian flood basalts." *Nature* 269:822–25.

Basu, A. R., M. I. Petaev, R. J. Poreda, S. B. Jacobsen, and L. Becker. 2003. "Chondritic meteorite fragments associated with the Permian-Triassic boundary in Antarctica." *Science* 302:1388–92.

Batten, R. L. 1973. "The vicissitudes of the Gastropoda during the interval of Guadelupian-Ladinian time," pp. 596–607 in A. Logan and L. V. Hills, eds. *The Permian and Triassic Systems and Their Mutual Boundary.* Canadian Society of Petroleum Geology, Calgary, Canada.

Baud, A., M. Magaritz, and W. T. Holser. 1989. "Permian-Triassic of the Tethys: carbon isotope studies." *Sonderdruck aus Geologische Rundschau* 78:649–77.

Baumiller, T. K., and F. J. Gahn. 2004. "Testing predator-driven evolution with Paleozoic crinoid arm regeneration." *Science* 305:1453–55.

Beauchamp, B. 1994. "Permian climatic cooling the Canadian Arctic," pp. 229–46 in G. D. Klein, ed. *Pangea: Paleoclimate, Tectonics and Sedimentation during Accretion, Zenith and Breakup of a Supercontinent.* GSA, Boulder, CO.

Beauchamp, B., and A. Baud. 2002. "Growth and demise of Permian biogenic chert along northwest Pangea: evidence for end-Permian collapse of thermohaline circulation." *Palaeogeography, Palaeoclimatology, Palaeoecology* 184:37–63.

Beauchamp, B., and A. Desrochers. 1997. "Permian warm- to very cold-water carbonates and cherts in northwest Pangea," in N. P. James, and J. A. Clarke, eds. *Cool Water Carbonates.* SEPM Special Publication No. 56, Pp. 327–47.

Becker, L., R. J. Poreda, A. G. Hunt, T. E. Bunch, and M. Rampino. 2001. "Impact event at the Permian-Triassic boundary: evidence from extraterrestrial noble gases in fullerenes." *Science* 291:1530–33.

Becker, L., R. J. Poreda, A. R. Basu, K. O. Pope, T. M. Harrison, C. Nicholson, and R. Iasky. 2004a. "Bedout: A possible end-Permian impact crater offshore of northwestern Australia." *Science* 304:1469–74.

———. 2004b. "Response to Comment by Glikson on 'Bedout: A possible end-Permian impact crater offshore of northwestern Australia.'" *Science* 306:613.

Becker, L., R. J. Poreda, and K. O. Pope. 2004c. Response to Comment by Wignall et al., on "Bedout: A possible end-Permian impact crater offshore of northwestern Australia." *Science* 306:613.

Becker, L., R. J. Poreda, A. R. Basu, K. O. Pope, T. M. Harrison, C. Nicholson, and R. Iasky. 2004d. "Response to Comment by Renne et al., on 'Bedout: A possible end-Permian impact crater offshore of northwestern Australia.' " *Science* 306:613.

Behrensmeyer, A. K., J. Damuth, W. A. DiMichele, R. Potts, H. D. Sues, and S. L. Wing. 1992. *Terrestrial Ecosystems through Time.* University of Chicago Press, Chicago.

Benton, M. J., V. P. Tverdokhlebov, and M. V. Surkov. 2004. "Ecosystem remodeling among vertebrates at the Permian-Triassic boundary in Russia." *Nature* 432:97–100.

Berner, R. A. 2002. "Examination of hypoheses for the Permo-Triassic boundary extinction by carbon cycle modeling." Proceedings of the National Academy of Sciences, USA 99:4172–77.

Bowring, S. A., D. H. Erwin, Y. G. Jin, M. W. Martin, K. L. Davidek, and W. Wang. 1998. "U/Pb zircon geochronology and tempo of the end-Permian mass extinction." *Science* 280:1039–45.

Brookfield, M. E., R. J. Twitchett, and C. Goodings. 2003. "Palaeoenvironments of the Permian-Triassic transitions sections in Kashmir, India." *Palaeogeography, Palaeoclimatology, Palaeoecology* 198:353–71.

Brown, R. D. 2002. "An impact centered near Cameroon at the time of the Permian extinction caused the fragmentation of Pangaea." *EOS Transactions of the American Geophysical Union* 83(19) Spring Meeting Supplement, Abstract T22A-08.

Buffett, B., and D. Archer. 2004. "Global inventory of methane clathrate: sensitivity to changes in the deep ocean." *Earth and Planetary Science Letters* 227:185–99.

Buseck, P. R. 2002. "Geological fullerenes: review and analysis." *Earth and Planetary Science Letters* 203:781–92.

Campbell, I. H., G. K. Czamanske, V. A. Fedorenko, R. I. Hill, and V. Stepanov. 1992. "Synchronism of the Siberian traps and the Permian-Triassic boundary." *Science* 258:1760–63.

Cao, C. Q., W. Wang, and J. Jin. 2002. "Carbon isotope excursions across the Permian-Triassic boundary in the Meishan section, Zhejiang Province, China." *Chinese Science Bulletin* 47:1125–30.

Carlson, S. J. 1991. "A phylogenetic perspective on articulate brachiopod diversity and the Permian Mass Extinction," pp. 119–42 in E. Dudley, ed. *The Unity of Evolutionary Biology. Proceedings of the 4th International Congress of Systematic and Evolutionary Biology.* Dioscorides Press, Portland, OR.

Chase, J. M., and M. A. Leibold. 2003. *Ecological Niches.* University of Chicago Press, Chicago, IL.

Chen, Z. Q., G. R. Shi, and K. Kaiho. 2002. "A new genus of rhynconellid brachiopod from the Lower Triassic of South China and implications for timing the recovery of Brachiopoda after the end-Permian mass extinction." *Palaeontology* 45:149–64.

Clark, D. J., C. Y. Wang, C. J. Orth, and J. S. Gilmore. 1986. "Conodont survival and low Iridium Anomalies: abundances across the Permian-Triassic boundary in South China." *Science* 233:984–86.

Courtillot, V., A. Davaille, J. Besse, and J. Stock. 2003. "Three distinct types of hotspots in the Earth's mantle." *Earth and Planetary Science Letters* 205: 295–308.

Courtillot, V., and P. R. Renne. 2003. "On the ages of flood basalt events." *Compte Rendu Geosciences* 335:113–40.

Czamanske, G. K., A. B. Gurevitch, V. A. Fedorenko, and O. Simonov. 1998. "Demise of the Siberian plume: paleogeographic and paleotectonic reconstruction from the prevolcanic and volcanic record, north-central Siberia." *International Geological Review* 40:95–115.

de Wit, M. J., J. G. Ghosh, S. de Villers, N. Rakotosolofo, J. Alexander, A. Tripathi, and C. V. Looy. 2002. "Multiple organic carbon isotope reversals across the Permo-Triassic boundary of Terrestrial Gondwana sequences: clues to extinction patterns and delayed ecosystem recovery." *Journal of Geology* 110:227–40.

Dickens, G. R., C. K. Paull, P. Wallace, and O.L.S. Party. 1997. "Direct measurement of in situ methane quantities in a large gas hydrate reservoir." *Nature* 385:427–28.

Dickens, G. R., P. Wallace, C. K. Paull, and W. S. Borowski. 2000. "Detection of methane gas hydrate in the pressure core sampler (PCS): volume-pressure-time relations during controlled degassing experiments," pp. 113–26 in C. K. Paull and R. Matsumoto, eds. *Proceedings of the Ocean Drilling Program, Scientific Results.* Ocean Drilling Program, College Station, Texas.

Elkins-Tanton, L. T., and B. H. Hager. 2000. "Melt intrusion as a trigger for lithospheric foundering and the eruption of the Siberian flood basalts." *Geophysical Research Letters* 27:3937–40.

Engoren, M. 2004. "Vertebrate extinction across Permo-Triassic boundary in Karoo Basin, South Africa: Discussion." *GSA Bulletin* 116:1294.

Erwin, D. H. 1993. *The Great Paleozoic Crisis: Life and Death in the Permian.* Columbia University Press, New York.

———. 1996. "Understanding biotic recoveries: extinction, survival and preservation during the end-Permian mass extinction," pp. 398–418 in D. Jablonski, D. H. Erwin, and J. H. Lipps, eds. *Evolutionary Paleobiology.* University of Chicago Press, Chicago.

———. 2001. "Lessons from the past: biotic recoveries from mass extinctions." *Proceedings of the National Academy of Sciences* 98:5399–5403.

Erwin, D. H., S. A. Bowring, and Y. G. Jin. 2002. "The End-Permian Mass Extinctions," in *Catastrophic Events and Mass Extinctions: Impacts and Beyond.* C. Koeberl and K. G. MacLeod, eds. Geological Society of America Special Paper 356: 363–83.

Erwin, D. H., J. W. Valentine, and J. J. Sepkoski, Jr. 1987. "A comparative study of diversification events: the early Paleozoic vs. the Mesozoic." *Evolution* 41:1177–86.

Eshet Y., M. R. Rampino and H. Visscher 1995. "Fungal event and palynological record of ecological crisis and recovery across the Permian-Triassic boundary." *Geology* 23: 967–70.

Ezaki, Y. 1997a. "Cold-water Permian rugosa and their extinction in Spitsbergen." *Bol. Royal Soc. Esp. Hist. Nat.* (Sec. Geol.) 92:381–88.

Ezaki, Y. 1997b. "Variations in the disappearance patterns of rugosan corals in Tethys and their implications for environments at the end of the Permian," pp. 126–33 in J. M. Dickins, Z. Y. Yang, H. F. Yin, S. G. Lucas, and S. K. Acharyya, eds. *Late Paleozoic and Early Mesozoic Circum-Pacific Events and Their Global Correlation.* Cambridge University Press, Cambridge.

Ezaki, Y., T. Kawamura, and K. Nakamura. 1994. "Kapp Starostin Formation in Spitsbergen: a sedimentary and faunal record of Late Permian paleoenvironments in an arctic region." *Canadian Society of Petroleum Geologists Memoir* 17:647–55.

Farley, K. A., S. Mukhopadhyay, Y. Isozaki, L. Becker, and R. J. Poreda. 2001. "An Extraterrestrial Impact at the Permian-Triassic Boundary?" *Science* 293:234a–g.

Fedorenko, V. A., and G. K. Czamanske. 1997. "Results of new field and geochemical studies of the volcanic and intrusive rocks of the Maymecha-Kotuy area, Siberian flood-Basalt Province, Russia." *International Geology Review* 39:479–51.

Fedorenko, V. A., G. K. Czamanske, T. Zen'ko, J. Budhan, and D. Siems. 2000. "Field and geochemical studies of the melilite-bearing Arydzhangsky Suite, and an overall perspective on the Siberian alkaline-ultramafic flood basalt volcanic rocks." *International Geology Review* 42:769–804.

Fedorowski, J. 1989. "Extinction of rugosa and tabulata near the Permian/Triassic boundary." *Acta Palaeontologica Polonica* 34:47–70.

Flügel, E. 2002. "Triassic reef patterns," in *Phanerozoic Reef Patterns, SEPM Special Publication No.* 72:391–462.

Flügel, E., and W. Kiessling. 2002. "Patterns of Phanerozoic reef crises." *Phanerozoic Reef Patterns, SEPM Special Publication No.* 72:691–733.

Flügel, E., and J. Reinhardt. 1989. "Uppermost Permian reefs in Skyros (Greece) and Sichuan (China): implications for the late Permian mass extinction event." *Palaios* 4:502–18.

Foote, M. 1995. "Morphological diversification of Paleozoic crinoids." *Paleobiology* 21:273–99.

———. 1996. "Ecological controls on the evolutionary recovery of post-Paleozoic crinoids." *Science* 274:1492–95.

Foster, C. A., G. A. Logan, and R. E. Summons. 1998. "The Permian-Triassic boundary in Australia: where is it and how is it expressed?" *Proc. R. Soc. Victoria* 110:247–66.

Foster, C. A., M. H. Stephenson, C. Marshall, G. A. Logan, and P. F. Greenwood. 2002. "A revision of *Reduviasponites* Wilson 1962: description, illustrations, comparison and biological affinities." *Palynology* 26:35–58.

Fraiser, M., and D. J. Bottjer. 2004. "The non-actualistic Early Triassic gastropod fauna: a case study of the Lower Triassic Sinbad Limestone Member." *Palaios* 19:259–75.

Girty, G. H. 1908. "The Guadalupian fauna." United States Geological Survey Professional Paper 58:1–651.

Glikson, A. 2003. "Comment on 'Extraterrestrial influences on mantle plume activity' by D. H. Abbott and A. E. Isley (2003)." *Earth and Planetary Science Letters* 215:425–27.

————. 2004. "Comment on 'Bedout: a possible end-Permian impact crater offshore of Northwestern Australia.' " *Science* 306:613. http://www.sciencemag .org/cgi/content/full/306/5696/613

Grabau, A. W. 1904. "Guide to the geology and paleontology of the Schoharie Valley in eastern New York." *58th Annual Report of the New York State Museum.* Vol. 3, Bulletin 92, *Paleontology* 13, pp. 77–386.

Gradstein, F. M., J. G. Ogg, A. G. Smith, F. P. Agterberg, W. Bleeker, R. A. Cooper, V. Davydov, P. Gibbard, L. Hinov, M. R. House, L. Lourens, H-P. Luterbacher, J. McArthur, M. J. Melchin, L. J. Robb, J. Shergold, M. Villeneuve, B. R. Wardlaw, J. Ali, H. Brinkhuis, F. J. Hilgen, J. Hooker, R. J. Howarth, A. H. Knoll, J. Lasker, S. Monechi, J. Powell, K. A. Plumb, I. Raffi, U. Rohl, A. Sanfilippo, B. Schmitz, N. J. Schakleton, G. A. Sheilds, H. Strauss, H. Van Dam, H. Veizer, Th. Van Kolfschoten, and D. Wilson. 2004. *A Geologic Time Scale 2004.* Cambridge University Press, Cambridge.

Grice, K., C. Q. Cao, G. D. Love, M. E. Böttcher, R. J. Twichett, E. Grosjean, R. E. Summons, S. C. Rurgeon, W. Dunning, and Y. G. Jin. 2005. "Photic zone euxinia during the Permian-Triassic superanoxia event." *Science* 307: 706–9.

Grotzinger, J. P., and A. H. Knoll. 1995. "Anomalous carbonate precipitates: is the Precambrian the key to the past?" *Palaios* 10:578–96.

Harland, W. B., A. G. Smith, and B. Wilcock. 1964. "Geological Society Phanerozoic time-scale 1964." *Quarterly Journal Geological Society* 120:260–62.

Harland, W. B., A. V. Cox, P. G. Llewellyn, C.A.G. Pickton, A. G. Smith, and R. Walters. 1982. *A Geologic Time Scale.* Cambridge University Press, Cambridge.

Harland, W. B., R. L. Armstrong, A. V. Cox, L. E. Craig, A. G. Smith, and D. G. Smith. 1989. *A Geologic Time Scale Scale 1989.* Cambridge University Press, Cambridge.

He, B., Y. G. Xu, S. L. Chung, L. Xiao, and Y. Wang. 2003. "Sedimentary evidence for a rapid, kilometer-scale crustal doming prior to the eruption of the Emeishan flood basalts." *Earth and Planetary Science Letters* 213:391–405.

Heydari, E., J. Hassanzadeh, W. J. Wade, and A. M. Ghazi. 2003. "Permian-Triassic boundary interval in the Abadeh section of Iran with implications for mass extinction: Part 1—Sedimentology." *Palaeogeography, Palaeoclimatology, Palaeoecology* 193:405–23.

Holser, W. T., and M. Magaritz. 1987. "Events near the Permian-Triassic boundary." *Modern Geology* 11:155–80.

Holser, W. T., H. P. Schonlaub, M. Attrep, Jr., K. Boeckelmann, P. Klein, M. Magaritz, C. J. Orth et al. 1989. "A unique geochemical record at the Permian/Triassic boundary." *Nature* 337:39–44.

Holser, W. T., H. P. Schonlaub, P. Klein, K. Boeckelmann, and M. Magaritz. 1991. "The Permian-Triassic boundary in the Gartnerkofel region of the Carnic Alps (Austria). Introduction." *Abh. Geol. B.-A* 45:5–16.

Hotinski, R. M., K. L. Bice, L. R. Kump, R. G. Najjar, and M. A. Arthur. 2001. "Ocean stagrantion and end-Permian anoxia." *Geology* 29:7–10.

Hotinski, R. M., L. R. Kump, and K. L. Bice. 2002. "Comment on 'Could the Late Permian deep ocean have been anoxic?' by R. Zhang et al.," *Paleoceanography* 17:1052–53.

Isozaki, Y. 1995. "Superanoxia across the Permo-Triassic boundary: record in accreted deep-sea pelagic chert in Japan." *Pangea: Global Environments and Research*. Canadian Society of Petroleum Geologists, Memoir 17:805–12.

———. 1997. "Permo-Triassic boundary superanoxia and stratified superocean: records from lost deep sea." *Science* 276:235–38.

Ivanov, B. A., and H. J. Melosh. 2003. "Impacts do not initiate volcanic eruptions: eruptions close to the crater." *Geology* 31:869–72.

Jablonski, D. 1986. "Causes and consequences of mass extinction: a comparative approach," pp. 183–229 in D. K. Elliot, ed. *Dynamics of Extinction*. John Wiley and Sons, New York.

Jin, Y. G., Q. H. Shang, and C. Q. Cao. 2000a. "Late Permian magnetostratigraphy and its global correlation." *Chinese Science Bulletin* 45:698–704.

Jin, Y. G., Y. Wang, W. Wang, Q. H. Shang, C. Q. Cao, and D. H. Erwin. 2000b. "Pattern of marine mass extinction near the Permian-Triassic boundary in South China." *Science* 289:432–36.

Jones, A., G. D. Price, N. J. Price, P. S. DeCarli, and R. A. Clegg. 2002. "Impact-induced melting and the development of large igneous provinces." *Earth and Planetary Science Letters* 202:551–61.

Kaiho, K., Y. Kajiwara, T. Nakano, Y. Miura, H. Kawahata, K. Tazaki, M. Ueshima, Z. Q. Chen, and G. R. Shi. 2001. "End-Permian catastrophe by a bolide impact: Evidence of a gigantic release of sulfur from the mantle." *Geology* 29:815–18.

Kajiwara, Y., S. Yamakita, K. Ishida, H. Ishiga, and A. Imai. 1994. "Development of a largely anoxic stratified ocean and its temporary mixing at the Permian/Triassic boundary supported by the sulfur isotope record." *Palaeogeography, Palaeoecology, Palaeoecology* 111:367–79.

Kakuwa, Y. 1996. "Permian-Triassic mass extinction event recorded in bedded chert sequence in southwest Japan." *Palaeogeography, Palaeoclimatology, Palaeoecology* 121:35–51.

Kakuwa, Y., and K. Toyoda. 1996. "Iridium anomaly is not detected at around the P/T boundary of bedded chert sequence in southwest Japan." *The Journal of the Geological Society of Japan* 102:139–42.

Kamo, S. L., G. K. Czamanske, Y. Amelin, V. A. Fedorenko, D. W. Davis, and V. R. Trofimov. 2003. "Rapid eruption of Siberian flood-volcanic rocks and evidence for coincidence with the Permian-Triassic boundary and mass extinction at 251 Ma." *Earth and Planet. Sci. Lett.* 214:75–91.

Kamo, S. L., G. K. Czamanske, Y. Amelin, V. A. Fedorenko, and V. R. Trofimov. 2000. "U-Pb zircon and Badellyite and U-Th-Pb perovskite ages from Siberian flood volcanism, Maymecha-Koruy area, Siberia." Goldschmidt 2000. Oxford,

UK. European Association for Geochemistry and the Geochemical Society, *Journal of Conference Abstracts* 5(2): 569.

Kamo, S. L., G. K. Czamanske, and T. E. Krough. 1996. "A minimum U-Pb age for Siberian flood-basalt volcanism." *Geochimica et Cosmochimica Acta* 60:3505–11.

Katz, M. E., D. K. Pak, G. R. Dickens, and K. G. Miller. 1999. "The source and fate of massive carbon input during the latest Paleocene thermal maximum." *Science* 286:1531–33.

Kennedy, M. J., D. R. Pevear, and R. J. Hill. 2002. "Mineral surface control of organic carbon in black shale." *Science* 295:657–60.

Kerr, R. A. 2003. "Has an impact done it again?" *Science* 302:1314–26.

———. 2004. "Evidence of huge, deadly impact found off Australian coast?" *Science* 304:491.

Kershaw, S., Guo, L., Swift, A., and Fan, J. S. 2002. "Microbialites in the Permian-Triassic boundary interval in central China: structure, age and distribution." *Facies* 47, 83–90.

Kershaw, S., T. Zhang, and G. Z. Lan. 1999. "A microbiolite carbonate crust at the Permian-Triassic boundary in South China and its palaeoenvironmental significance." *Palaeogeography, Palaeoclimatology, Palaeoecology* 146:1–18.

Kidder, D. L., and T. R. Worseley. 2004. "Causes and consequences of extreme Permo-Triassic warming to globally equable climate and relation to the Permo-Triassic extinction and recovery." *Palaeogeography, Palaeoclimatology, Palaeoecology* 203:207–37.

Knoll, A. H. 1984. "Patterns of extinction in the fossil record of vascular plants," pp. 23–68 in M. H. Nitecki, ed. *Extinction.* University of Chicago Press, Chicago.

Knoll, A. H., R. K. Bambach, D. E. Canfield, and J. P. Grotzinger. 1996. "Comparative earth history and late Permian mass extinction." *Science* 273:452–57.

Koeberl, C., K. A. Farley, B. Peucker-Ehrenbrink, and M. A. Sephton. 2004. "Geochemistry of the end-Permian extinction event in Austia and Italy: no evidence for an extraterrestrial component." *Geology* 30:1053–56.

Korte, C., H. Kozur, M. M. Joachimski, H. Strauss, J. Veizer, and L. Schwark. 2004. "Carbon, sulfur, oxygen and strontium isotope records, organic geochemistry and biostratigraphy across the Permian/Triassic boundary in Abadeh, Iran." *International Journal of Earth Sciences (Geol. Rundsch)* 93:565–81.

Kozur, H. 1996. "The conodonts *Hindeodus, Isarciella* and *Sweetohindeodus* in the uppermost Permian and lowermost Triassic." *Geol. Croat* 49:81–115.

Krull, E. S., and G. J. Retallack. 2000. "Del 13C depth profiles from paleosols across the Permian-Triassic boundary: evidence for methane release." *GSA Bulletin* 112(9): 1459–72.

Krull, E. S., G. J. Retallack, H. J. Campbell, and G. L. Lyon. 2000. "Del $^{13}C_{org}$ chemostratigraphy of the Permian-Triassic boundary in the Maitai Group, New Zealand: evidence for high-latitudinal methane release." *New Zealand Journal of Geology and Geophysics* 43:21–32.

Krystyn, L., S. Richoz, A. Baud, and R. J. Twitchett. 2003. "A unique Permian-Triassic boundary section from the Neotethyan Hawasina Basin, Central Oman Mountains." *Palaeogeography, Palaeoclimatology, Palaeoecology* 191:329–44.

Labandeira, C. C. 2005. "The fossil record of insect extinction: new approaches and future directions." *American Entomologist* 51:14–29.

Labandeira, C. C., and J. J. Sepkoski, Jr. 1993. "Insect diversity in the fossil record." *Science* 261:310–15.

Laland, K. N., F. J. Odling-Smee, and M. W. Feldman. 1999. "Evolutionary consequences of niche construction and their implications for ecology." *Proceedings National Academy Sciences, USA* 96:10242–47.

————. 2000. "Niche construction, biological evolution and cultural change." *Behavioral and Brain Sciences* 23:131–75.

Lehrmann, D. J. 1999. "Early Triassic calcimicrobial mounds and biostromes of the Nanpanjiang basin, South China." *Geology* 27:359–62.

Lehrmann, D. J., J. L. Payne, S. V. Felix, P. M. Dillett, H. M. Wang, Y. Y. Yu, and J. Y. Wei. 2003. "Permian-Triassic boundary sections from the shallow-marine carbonate platforms of the Nanpanjiang Basin, South China: implications for oceanic conditions associated with the end-Permian extinction and its aftermath." *Palaios* 18:138–52.

Lind, E. N., S. V. Kropotov, G. K. Czamanske, S. C. Gromme, and V. A. Fedorenko. 1994. "Paleomagnetism of the Siberian flood basalts of the Noril'sk area: a constraint on eruption duration." *International Geology Review* 36:1139–50.

Lo, C. H., S. L. Chung, T. Y. Lee, and G. Y. Wu. 2002. "Age of the Emeishan flood magmatism and relations to Permian-Triassic boundary events." *Earth and Planetary Science Letters* 198:449–58.

Looy, C. V., Brugman, W. A., Dilcher, D. L. and Visscher, H. 1999. "The delayed resurgence of equitorial forests after the Permian-Triassic ecologic crisis." *Proceedings of the National Academy of Sciences, USA* 96:13857–862.

Looy, C. V., Twitchett, R. J., Dilcher, D. L., Van Konijnenburg-Van Cittert, J.H.A., and Visscher, H. 2001. "Life in the end-Permian dead zone." *Proceedings of the National Academy of Sciences, USA* 98:7879–83.

MacArthur, R. H., and E. O. Wilson. 1967. *The Theory of Island Biogeography.* Princeton University Press, Princeton, NJ.

MacLeod, K. G., R.M.H. Smith, P. L. Koch, and P. D. Ward. 2000. "Timing of mammal-like reptile extinctions across the Permian-Triassic boundary in South Africa." *Geology* 28:227–30.

Magaritz, M., R. V. Krishnamurthy, and W. T. Holser. 1992. "Parallel trends in organic and inorganic isotopes across the Permian/Triassic boundary." *American Journal of Science* 292:727–39.

Marshall, C. R. 1994. "Confidence intervals on stratigraphic ranges: partial relaxation of the assumption of randomly distributed fossil horizons." *Paleobiology* 20:459–69.

Marshall, C. R., and P. D. Ward. 1996. "Sudden and gradual molluscan extinctions in the latest Cretaceous of western European Tethys." *Science* 274:1360–63.

Martin, E. E., and J. D. Macdougall. 1995. "Sr and Nd isotopes at the Permian/ Triassic boundary: a record of climate change." *Chemical Geology* 125:73–95.

Maruoka, T., C. Koeberl, P. J. Hancox, and W. U. Reimold. 2003. "Sulfur geochemistry across a terrestrial Permian-Triassic boundary section in the Karoo Basin, South Africa." *Earth and Planetary Science Letters* 206:101–17.

McGhee, G. R., Jr., P. M. Sheehan, D. J. Bottjer, and M. L. Droser. 2004. "Ecological ranking of Phanerozoic biodiversity crises: ecological and taxonomic severities are decoupled." *Palaeogeography, Palaeoclimatology, Palaeoecology* 211:289–97.

McGowan, A. J. 2004a. "Ammonoid taxonomic and morphologic recovery patterns after the Permian-Triassic." *Geology* 32:665–68.

———. 2004. "The effects of the Permo-Triassic bottleneck on Triassic ammonoid morphological evolution." *Paleobiology* 30:369–95.

McLoughlin, S., S. Lindstrom, and A. N. Drinnan. 1997. "Gondwanan floristic and sedimentological trends during the Permian-Triassic transition: new evidence from the Amery group, northern Prince Charles Mountains, East Antarctica." *Antarctic Science* 9:281–98.

McRoberts, C. A. 2001. "Triassic bivalves and the initial marine Mesozoic revolution: a role for predators." *Geology* 29:359–62.

Metcalfe, I., R. S. Nicoll, R. Mundil, C. Foster, J. Glen, J. Lyons, X. F. Wang, C. Y. Want, P. R. Renne, L. Black, Q. Xun, and X. D. Mao. 2001. "The Permian-Triassic boundary and mass extinction in China." *Episodes* 24:239–44.

Michaelsen, P. 2002. "Mass extinction of peat-forming plants and the effect of fluvial styles across the Permian-Triassic boundary, northern Bowen Basin, Australia." *Palaeogeography, Palaeoclimatology, Palaeoecology* 179:178–88.

Milkov, A. B., G. E. Claypool, J. Y. Lee, W. Y. Xu, G. R. Dickens, W. S. Borowski, and O.L.S. Party. 2003. "In situ methane concentrations at Hydrate Ridge, offshore Oregon: new constraints on the global gas hydrate inventory from an active margin." *Geology* 31:833–36.

Modesto, S., R. Damiani, J. Neveling, and A. Yates. 2003. "A new Triassic owenettid pararreptile and the mother of mass extinctions." *Journal of Vertebrate Paleontology* 23:715–19.

Modesto, S., H. D. Sues, and R. Damiani. 2001. "A new Triassic procolophonoid reptile and its implications for Procolophonid survivorship during the Permo-Triassic extinction event." *Proceedings of the Royal Society London*, Series B 268:2047–52.

Morgan, J. P., T. J. Reston, and C. R. Ranero. 2004. "Contemporaneous mass extinctions continental flood basalts, and 'impact signals': are mantle-plume-induced lithospheric gas explosions the causal link?" *Earth and Planetary Science Letters* 217:263–84.

Mory, A. J., R. P. Lasky, A. Y. Glikson, and F. Pirajno. 2000. "Woodleigh, Carnavon Basin, Western Australia: a new 120-km-diameter impact structure." *Earth and Planetary Science Letters* 177:118–28.

———. 2000. "Response to 'Critical comment on A. J. Mory et al., Woodleigh Carnarovan Basin, Western Australia: a new 120 km diameter impact structure.' " *Earth and Planetary Science Letters* 184:359–65.

Mundil, R., K. R. Ludwig, I. Metcalfe, and P. R. Renne. 2004. "Age and timing of the end Permian mass extinctions: U/Pb geochronology on closed-system zircons." *Science* 305:1760–63.

Mundil, R., I. Metcalfe, K. R. Ludwig, P. R. Renne, F. Oberli, and R. S. Nicoll. 2001. "Timing of the Permian-Triassic biotic crisis: implications from new zircon U/Pb age data (and their limitations)." *Earth and Planetary Science Letters* 187:131–45.

Musashi, M., Y. Isozaki, T. Koike, and R. Krueulen. 2001. "Stable carbon isotope signature in mid-Panthalassa shallow-water carbonates across the Permo-Triassic boundary: evidence for ^{13}C-depleted superocean." *Earth and Planetary Science Letters* 191:9–20.

Newell, A. J., V. P. Tverdokhlebov, and M. J. Benton. 1999. "Interplay of tectonics and climate on a transverse fluvial system, Upper Permian, southern Uralian foreland Basin, Russia." *Sedimentary Geology* 127:11–29.

Newell, N. D., J. K. Rigby et al. 1953. *The Permian Reef Complex of the Guadalupe Mountains Region, Texas and New Mexico.* W. H. Freeman, San Francisco.

Newhall, C. G., J. A. Power, and R. S. Punongbayan. 2002. "To make grow." *Science* 295:1241–42.

Nielsen, J. K., and Y. Shen. 2004. "Evidence for sulfidic deep water during the Late Permian in the East Greenland Basin." *Geology* 32:1037–40.

Nikishin, A. M., P. Ziegler, A. D. Abbott, M. Brunet, and S. Cloetingh. 2002. "Permo-Triassic intraplate magmatism and rifting in Eurasia: implications for mantle plumes and mantle dynamics." *Tectonophysics* 351:2–39.

Odling-Smee, F. J., K. N. Laland, and M. W. Feldman. 2003. *Niche Construction.* Princeton University Press, Princeton, NJ.

Orchard, M. J., and L. Krystyn. 1998. "Conodonts of the lowermost Triassic of Spiti, and a new zonation based on *Neogondolella* successions." *Rivista Italiana di Paleontologia e Stratigrafia* 104:341–68.

Orth, C. J., and L. R. Quintana. 1991. "The Permian-Triassic of the Gartnerkofell core (Carnic Alps, Austria): geochemistry of common and trace elements II—INAA and RNAA." *Abhandlungen der Geologishen Bundesanstalt* 45:79–98.

Palmer, E., 1966. *The Plains of Camdeboo.* Collins, London.

Paul, J., and T. M. Peryt. 2000. "Kalkowsky's stromatolites revisited (Lower Triassic Bundsandstein, Harz Mountains, Germany)." *Palaeogeography, Palaeoclimatology, Palaeoecology* 161:435–58.

Petaev, M. I., S. B. Jacobsen, A. R. Basu, and L. Becker. 2004. "Magnetic Fe, Si, Al-rich impact sperules from the P-t boundary layer at Graphite Peak, Antarctica." *Lunar and Planetary Science Conference,* Abstracts.

Poreda, R. J., and L. Becker. 2003. "Fullerenes and interplanetary dust particles at the Permian-Triassic boundary." *Astrobiology* 3:75–90.

Pruss, S., and D. J. Bottjer. 2004. "Late Early Triassic microbial reefs of the western United States: a description and model for their deposition in the aftermath of the end-Permian mass extinction." *Palaeogeography, Palaeoclimatology, Palaeoecology,* 211:127–37.

Pruss, S., M. Frasier, and D. J. Bottjer. 2004. "Proliferation of Early Triassic wrinkle structures: implications for environmental stress following the end-Permian mass extinction." *Geology* 32:461–64.

Puffer, J. H. 2001. "Contrasting high-field strength-element contents of continental flood basalts from plume versus reactivated-arc sources." *Geology* 29:675–78.

Racki, G. 2003. "End-Permian mass extinction: oceanographic consequences of double catastrophic volcanism." *Lethaia* 36:171–73.

Rampino, M. R., and A. C. Adler. 1998. "Evidence for abrupt latest Permian mass extinction of foraminifera: results of tests for the Signor-Lipps effect." *Geology* 26:415–18.

Rampino, M. R., A. Prokoph, and A. Adler. 2000. "Tempo of the end-Permian event: high-resolution cyclostratigraphy at the Permian-Triassic boundary." *Geology* 28:643–46.

Raup, D. M. 1991. *Extinction: Bad Genes or Bad Luck?* Norton, New York.

Raup, D., and G. Boyajian. 1988. "Patterns of generic extinction in the fossil record." *Paleobiology* 14:109–25.

Raup, D. M., and J. J. Sepkoski, Jr. 1984. "Periodicity of extinction in the geologic past." *Proceedings of the National Academy of the Sciences, USA*, 81:801–805.

———. 1986. "Periodic extinction of families and genera." *Science* 231:833–36.

Ray, S., and A. Chinsamy. 2003. "Fuctional aspects of the post-cranial anatomy of the Permian dicynodont *Diictodon* and their ecological implications." *Palaeontology* 46:131–83.

Rees, P. M. 2002. "Land-plant diversity and the end-Permian mass extinction." *Geology* 30:827–30.

Reichow, M. K., A. D. Saunders, R. V. White, M. S. Pringle, A. I. Al'Mukhamedov, A. I. Medvedev, and N. P. Kirda. 2002. "^{40}Ar/^{39}Ar dates from the West Siberian Basin: Siberian flood basalt province doubled." *Science* 296:1846–49.

Reichow, M. K., A. D. Saunders, R. V. White, A. I. Al'Mukhamedov, and A. Y. Medvedev. 2005. "Geochemistry and petrogenesis of basalts from the West Siberian Basin: an extension of the Permo-Triassic Siberian Traps, Russia." *Lithos* 79:425–52.

Reimold, W. U., V. von Brunn, and C. Koeberl. 1997. "Are diamictites impact ejecta? No supporting evidence from South African Dwyka group diamictite." *Journal of Geology* 105:517–30.

Remane, J. 2000. International Stratigraphic Chart, with explanatory note. Sponsored by ICS, IUGS and UNESCO. 31st International Geological Congress, Rio de Janeiro 2000, p. 16.

Renne, P. R., and A. R. Basu. 1991. "Rapid eruption of the Siberian Traps flood basalts at the Permo-Triassic boundary." *Science* 253:176–79.

Renne, P. R., H. J. Melosh, K. A. Farley, U. W. Reimold, C. Koeberl, M. R. Rampino, S. P. Kelly, B. A. Ivanov. 2004. "Is Bedout an impact crater? Take 2." *Science* 306:613.

Renne, P. R., Z. C. Zhang, M. A. Richards, M. T. Black, and A. R. Basu. 1995. "Synchrony and causal relations between Permian-Triassic boundary crises and Siberian flood volcanism." *Science* 269:1413–16.

Retallack, G. J. 1995. "Permian-Triassic life crisis on land." *Science* 267:77–80.

———. 1997. "Earliest Triassic origin of *Isoetes* and quillwort evolutionary radiation." *Journal of Paleonotology* 71:500–21.

——— 1999. "Postapocalyptic greenhouse paleoclimate revealed by earliest Triassic paleosols in the Sidney Basin, Australia." *GSA Bulletin* 111:52–70.

———. 2004. "Vertebrate extinction across Permo-Triassic boundary in Karoo Basin, South Africa: Reply." *GSA Bulletin* 116:1295–96.

Retallack, G. J., A. Seyedolai, E. S. Krull, W. T. Holser, C. P. Ambers, and F. T. Kyte. 1998. "Search for evidence of impact at the Permian-Triassic boundary in Antarctica and Australia." *Geology* 26:979–82.

Retallack, G. J., and E. S. Krull. 1999. "Landscape ecological shift at the Permian-Triassic boundary in Antarctica." *Australian Journal of Earth Sciences* 46:785–812.

Retallack, G. J., R.M.H. Smith, and P. D. Ward. 2003. "Vertebrate extinction across Permian-Triassic boundary in Karoo Basin, South Africa." *GSA Bulletin* 115:1133–52.

Retallack, G. J., J. J. Veevers, and R. Morante. 1996. "Global coal gap between Permian-Triassic extinction and Middle Triassic recovery of peat-forming plants." *GSA Bulletin* 108:195–207.

Rigby, J. K., and A. B. Millward 1988. "A look back at the Permian Reefs of West Texas and New Mexico." *Earth Sciences History* 7:71–89.

Rodland, D. L., and D. J. Bottjer. 2001. "Biotic recovery from the end-Permian mass extinction: behavior of the inarticulate brachiopod *Lingula* as a disaster taxon." *Palaios* 16:95–101.

Rohl, U., et al. 2000. "New chronology for the late Paleocene thermal maximum and its environmental interpretation." *Geology* 28:927–30.

Roy, K. 1994. "Effects of the Mesozoic marine revolution on the taxonomic, morphologic and biogeographic evolution of a group: aporrhaid gastropods during the Mesozoic." *Paleobiology* 20:274–96.

Rubidge, B. S. 1995. "Biostratigraphy of the Beaufort Group (Karoo Supergroup)." *Geological Survey of South Africa*, Biostratigraphic Series, No. 1:1–46.

Rubidge, B. S., and C. A. Sidor. 2001. "Evolutionary patterns among Permo-Triassic therapsids." *Annual Review of Ecology and Systematics* 32:449–80.

Ryskin, G. 2003. "Methane-driven oceanic eruptions and mass extinctions." *Geology* 31:741–44.

Sano, H., and K. Nakashima. 1997. "Lowermost Triassic (Griesbachian) microbial bindstone-cementstone facies, southwest Japan." *Facies* 36:1–24.

Sarkar, A., H. Yoshioka, M. Ebihara, and H. Naraoka. 2003. "Geochemical and organic carbon isotope studies across the continental Permo-Triassic boundary of Raniganj Basin, eastern India." *Palaeogeography, Palaeoclimatology, Palaeoecology* 191:1–14.

Saunders, A. D., R. W. England, M. K. Reichow, and R. V. White. 2005. "A mantle plume origin for the Siberian traps: uplift and extension in the West Siberian Basin, Russia." *Lithos* 79:407–24.

Schopf, T.J.M. 1974. "Permo-Triassic extinctions: relation to sea-floor spreading." *Journal of Geology* 82:129–43.

Schubert, J. K., and D. J. Bottjer. 1995. "Aftermath of the Permian-Triassic mass extinction event: paleoecology of Lower Triassic carbonates in the western USA." *Palaeogeography, Palaeoclimatology, Palaeoecology* 116:1–39.

Scotese, C.R., and R. P. Langford, 1995. "Pangea and the Palegeography of the Permian," pp.3–19 in P. A. Scholle, T. M. Peryt, and D. S. Ulmer-Scholle, eds. *The Permian of Northern Pangea.* Vol. 1, *Paleogeography, Paleoclimates, and Stratigraphy.* Springer-Verlag, Berlin.

Senowbari-Daryan, B., R. Zuhlke, T. Bechstadt, and E. Flugel. 1993. "Anisian (Middle Triassic) buildups of the northern dolomites (Italy): the recovery of reef communities after the Permian-Triassic crisis." *Facies* 28:181–265.

Sephton, M. A., C. V. Looy, R. J. Veefking, H. Brinkhuis, J. W. de Leeuw, and H. Visscher. 2002. "Synchronous record of the del 13C shifts in the oceans and atmosphere at the end of the Permian," pp. 455–62 in C. Koeberl and K. G. MacLeod, eds. *Catastrophic Events and Mass Extinctions: Impacts and Beyond.* Geological Society of America, Boulder, CO.

Sepkoski, J. J., Jr. 1979. "A kinetic model of Phanerozoic taxonomic diversity II. Early Paleozoic families and multiple equilibria." *Paleobiology* 5:222–51.

———. 1981. "A factor analytic description of the Phanerozoic marine fossil record." *Paleobiology* 7:36–53.

———. 1982. "A compendium of fossil marine families." Milwaukee Public Museum Contributions in *Biology and Geology* 51:125.

———. 1984. "A kinetic model of Phanerozoic taxonomic diversity. III. Post-Paleozoic families and mass extinction." *Paleobiology* 10:246–67.

———. 1987. "Environmental trends in extinction during the Paleozoic." *Science* 235:64–65.

———. 1991. "A model of onshore-offshore change in faunal diversity." *Paleobiology* 17:58–77.

———. 1997. "Biodiversity: past, present, and future." *Journal of Paleontology* 71:533–39.

Sepkoski, J. J., Jr., and A. I. Miller. 1985. "Evolutionary marine faunas and the distribution of Paleozoic benthic communities in space and time," pp. 153–89 in J. W. Valentine, ed. *Phanerozoic Diversity Patterns.* Princeton University Press, Princeton, NJ.

Sheldon, N. D., and G. J. Retallack. 2002. "Low oxygen levels in earliest Triassic soils." *Geology* 30:919–22.

Shen, J. W., T. Kawamura, and W. R. Yang. 1998. "Upper Permian coral reef and colonial rugose corals in northwest Hunan, South China." *Facies* 39:35–66.

Shen, S. Z., N. W. Archbold, and G. R. Shi. 2000. "Changhsingian (Late Permian) brachiopod palaeobiogeography." *Historical Biology* 15:121–34.

Shen, S. Z., and G. R. Shi. 2002. "Paleobiogeographical extinction patterns of Permian brachiopods in the Asian-western Pacific region." *Paleobiology* 28:449–63.

Shi, G. R., and S. Z. Shen. 2000. "Asian-western Pacific Permian brachiopoda in space and time: biogeography and extinction patterns," pp. 327–52 in H. F. Yin, J. M. Dickins, G. R. Shi and J. N. Tong, eds. *Permian-Triassic Evolution of Tethys and Western Circum-Pacific.* Elsevier, Amsterdam.

Shoemaker, E. M. 1983. "Asteroid and comet bombardment of the Earth." *Annual Review of Earth and Planetary Sciences* 11:461–94.

Signor, P. W., and C. E. Brett. 1984. "The mid-Paleozoic precursor to the Mesozoic marine revolution." *Paleobiology* 10:229–45.

Signor, P. W., III, and J. H. Lipps. 1982. "Sampling bias, gradual extinction patterns and catastrophes in the fossil record," pp. 283–90 in L. T. Silver and P. H. Schultz, eds. *Geological Implications of Impact Hypothesis of Large Asteroids and Comets on the Earth,* Special Paper. Geological Society of America, Boulder, CO.

Simberloff, D. S. 1974. "Permian mass extinction: effects of area on biotic equilibrium." *Journal of Geology* 82:267–74.

Smith, R.M.H., and P. D. Ward. 2001. "Pattern of vertebrate extinctions across an event bed at the Permian-Triassic boundary in the Karoo Basin of South Africa." *Geology* 29:1147–50.

Stanley, G. D., Jr. 2003. "The evolution of modern corals and their early history." *Earth-Science Reviews* 60:195–225.

Stanley, S. M. 1988a. "Paleozoic mass extinction: shared patterns suggest global cooling as a common cause." *American Journal of Science* 288:334–52.

———. 1988b. "Climatic cooling and mass extinction of Paleozoic reef communities." *Palaios* 3:228–32.

Stanley, S. M., and X. Yang. 1994. "A double mass extinction at the end of the Paleozoic era." *Science* 266:1340–44.

Steiner, M. B., Y. Eshet, M. R. Rampino, and D. M. Schwindt. 2003. "Fungal abundance spike and the Permian-Triassic boundary in the Karoo supergroup (South Africa)." *Palaeogeography, Palaeoclimatology, Palaeoecology* 194:405–14.

Svensen, H., S. Planke, A. Malthe-Sorenssen, B. Jamtveit, R. Myklebust, T. R. Eidem, and S. S. Rey. 2004. "Release of methane from a volcanic basin as a mechanism for initial Eocene global warming." *Nature* 429:542–45.

Sweet, W. C. 1992. "A conodont-based high-resolution biostratigraphy for the Permo-Triassic boundary interval," pp. 120–33 in W. C. Sweet, Z. Y. Yang, J. M. Dickins, and H. F. Yin, eds. *Permo-Triassic Boundary Events in the Eastern Tethys.* Cambridge University Press, Cambridge.

Szurlies, M., G. H. Bachmann, M. Menning, N. R. Nowaczyk, and K. C. Kading. 2003. "Magnetostratigraphy and high-resolution lithostratigraph of the Permian-Triassic boundary interval in Central Germany." *Earth and Planetary Science Letters* 212:263–78.

Tanner, L., S. G. Lucas, and M. Chapman. 2004. "Assessing the record and causes of Late Triassic extinctions." *Earth-Science Reviews* 65:103–39.

Tong, J. N., H. F. Yin, and Y. G. Jin. 1999. "Permian and Triassic sequence stratigraphy and sea level changes of eastern Yangtze platform." *Journal of the China University of Geosciences* 10:161–69.

Tverdokhlebov, V. P., G. I. Tverdokhlebova, A. V. Minikh, M. V. Surkov, and M. J. Benton. 2005. "Upper Permian vertebrates and their sedimentological context in the southern Urals, Russia." *Earth-Science Reviews* 69:27–77.

Twitchett, R. J. 1999. "Palaeoenvironments and faunal recovery after the end-Permian mass extinction." *Palaeogeography, Palaeoclimatology, Palaeoecology* 154:27–37.

Twitchett, R. J., L. Krystyn, A. Baud, J. R. Wheeley, and S. Richoz. 2004. "Rapid marine recovery after the end-Permian mass-extinction event in the absence of marine anoxia." *Geology* 32:805–808.

Twitchett, R. J., C. V. Looy, R. Morante, H. Visscher, and P. Wignall. 2001. "Rapid and synchronous collapse of marine and terrestrial ecosystems during the end-Permian biotic crisis." *Geology* 29:351–54.

Twitchett, R. J., and P. Wignall. 1996. "Trace fossils and the aftermath of the Permo-Triassic mass extinction: evidence from northern Italy." *Palaeogeography, Palaeoclimatology, Palaeoecology* 124:137–51.

Valentine, J. W., and E. M. Moores. 1970. "Plate-tectonic regulation of faunal diversity and sea level: a model." *Nature* 228:657–59.

———. 1973. "Provinciality and diversity across the Permian-Triassic boundary," pp. 759–66 in A. Logan and L. V. Hills, eds. *The Permian and Triassic Systems and Their Mutual Boundary.* Canadian Society of Petroleum Geologists, Calgary, Canada.

Veevers, J. J., P. J. Conaghan, and S. E. Shaw. 1994. "Turning point in Pangean environmental history at the Permian/Triassic (P/Tr) boundary," pp. 187–96 in G. Klein, ed. *Pangea: Paleoclimate, Tectonics, and Sedimentation during Accretion, Zenith and Breakup of a Supercontinent, Special Paper 288.* Geological Society of America, Boulder, CO.

Vermeij, G. J. 1977. "The Mesozoic marine revolution: Evidence from snails, predators and grazers." *Paleobiology* 3:245–58.

———. 1987. *Evolution and Escalation: An Ecologic History of Life.* Princeton University Press, Princeton, NJ.

Visscher, H., H. Brinkhuis, D. L. Dilcher, W. C. Elsik, Y. Eshet, C. V. Looy, M. R. Rampino, and A. Traverse. 1996. "The terminal Paleozoic fungal event: evidence of terrestrial ecosystem destabilization and collapse." *Proceedings of the National Academy of Sciences, USA* 93:2155–58.

Wang, Z. Q. 1996. "Recovery of vegetation from the terminal Permian mass extinction in North China." *Review of Paleobotany and Palynology* 91:121–42.

Wang, X. D., and T. Sugiyama. 2000. "Diversity and extinction patterns of Permian coral faunas of China." *Lethaia* 33:285–94.

Wanner, J. 1930. "Neue Beitrag zur Kenntnis der Permischen Echinodermen von Timor." *Wetenschappelijke Mededeelingen* 14:1–60.

———. 1931. "Neue Beitrage zur Kentnis der Permischen Echinodermen von Timor V. Poteriocrinidase, VI Blastoidea." *Wetenschappelijke Mededeelingen* 16:1–76.

———. 1937. "Neue Beitrage zur Kentnis der Permischen Echinodermen von Timor VIII–XIII. Beitrage zur Geologie von Niederlandischen-Indien. "*Sonder-Abduk aus Palaeontographica* Suppl. IV: 59–212.

Ward, P. D., J. Botha, R. Buick, M. O. De Kock, D. H. Erwin, G. Garrison, J. L. Kirschvink, and R. Smith. 2005. "Abrupt and gradual extinction among Late Permian land vertebrates in the Karoo Basin, South Africa." *Science,* 307:709–14.

Ward, P. D., D. R. Montgomery, and R. Smith. 2000. "Altered river morphology in South Africa related to the Permian-Triassic extinction." *Science* 289:1740–43.

Weidlich, O. 2002. "Middle and Late Permian reefs—distributional patterns and reservoir potential," pp. 339–90. *Phanerozoic Reef Patterns.* SEPM Special Publication.

Weidlich, O., W. Kiessling, and E. Flugel. 2003. "Permian-Traissic boundary interval as a model for forcing marine ecosystem collapse by long-term atmospheric oxygen drop." *Geology* 31:961–64.

Wignall, P. B. 2001. "Large igneous provinces and mass extinction." *Earth-Science Reviews* 53:1–33.

Wignall, P. B., and M. J. Benton, 1999. "Lazarus taxa and fossil abundance at times of biotic crisis." *Journal of the Geological Society London,* 156:453–56.

Wignall, P. B., and A. Hallam. 1992. "Anoxia as a cause of the Permo-Triassic mass extinction: facies evidence from northern Italy and the western United States." *Palaeogeography, Palaeoclimatology, Palaeoecology* 93:21–46.

———. 1993. "Griesbachian (earliest Triassic) palaeoenvironmental changes in the Salt Range, Pakistan and southeast China and their bearing on the Permo-Triassic mass extinction." *Palaeogeography, Palaeoclimatology, Palaeoecology* 102:215–37.

Wignall, P. B., A. Hallam, X. L. Lai, and F. Q. Yang. 1995. "Palaeoenvironmental changes across the Permian/Triassic boundary at Shangsi (N. Sichuan, China)." *Historical Biology* 10:175–89.

Wignall, P. B., R. Morante, and R. Newton. 1998. "The Permo-Triassic transition in Spitsbergen: $^{13}C^{org}$ chemostratigraphy, Fe and S geochemistry, facies, fauna and trace fossils." *Geological Magazine* 135:47–62.

Wignall, P. B., and R. Newton. 2003. "Contrasting deep-water records from the Upper Permian and Lower Triassic of south Tibet and British Columbia: evidence for a diachronous mass extinction." *Palaios* 18:153–67.

Wignall, P. B., R. Newton, and M. E. Brookfield. 2005. "Pyrite framboid evidence for oxygen-poor deposition during the Permian-Triassic crisis in Kashmir." *Palaeogeography, Palaeoclimatology, Palaeoecology* 216:183–88.

Wignall, P. B., B. Thomas, R. Willink, and J. Watling, 2004. "Is Bedout an impact crater? Take 1." *Science* 304:613.

Wignall, P. B., and R. J. Twitchett. 1996. "Oceanic anoxia and the end Permian mass extinction." *Science* 272:1155–58.

Wood, R. 1999. *Reef Evolution.* Oxford University Press, Oxford.

Wood, R., J.A.D. Dickson, and B. Kirkland-George. 1994. "Turning the Capitan reef upside down: a new appraisal of the ecology of the Permian Capitan Reef, Guadalupe Mountains, Texas and New Mexico." *Palaios* 9:422–27.

Woods, A. D., and D. J. Bottjer. 2000. "Distribution of ammonoids in the Lower Triassic Union Wash Formation (Eastern California): evidence for paleoceanographic conditions during recovery from the end-Permian mass extinction." *Palaios* 15:535–45.

Woods, A. D., D. J. Bottjer, M. Mutti, and J. Morrison. 1999. "Lower Triassic large sea-floor carbonates cements: their origin and a mechanism for the prolonged biotic recovery from the end-Permian mass extinction." *Geology* 27:645–48.

Xu, D. Y., S. L. Ma, Z. F. Chai, X. Y. Mao, Y. Y. Sun, Q. W. Zhang, and Z. Z. Yang. 1985. "Abundance variation of Iridium anomolies and trace elements at the Permian/Triassic boundary at Shagsi in China." *Nature* 314:154–56.

Xu, Y. G., B. He, S. L. Chung, M. A. Menzies, and F. A. Frey. 2004. "Geologic, geochemical and geophysical consequences of plume involvement in the Emeishan flood basalt province." *Geology* 32:917–20.

Zhang, R., M. Follows, J. P. Grotzinger, and J. Marshall. 2001. "Could the late Permian ocean have been anoxic?" *Paleooceanography* 16:317–29.

Zhang, R., M. J. Follows, and J. Marshall. 2003. "Reply to Comment by Roberta M. Hotinski, Lee R. Kump, and Karen L. Bice on 'Could the Late Permian deep ocean have been anoxic?' " *Earth and Planetary Science Letters* 18:Art No. 1015.

Zhou, M. F., J. Malpas, X. Y. Song, P. T. Robinson, M. Sun, A. K. Kennedy, C. M. Lesher, and R. R. Keays. 2002. "A temporal link between the Emeishan large igneous province (SW China) and the end-Guadalupian mass extinction." *Earth and Planetary Science Letters* 196:113–22.

Ziegler, A. M., and W. Charles. 1990. "Phytogeographic patterns and continental configurations during the Permian Period," pp. 363–79 in W. S. McKerrow and C. R. Scotese, eds. *Palaeozoic Palaeogeography and Biogeography.* Geologica Society Memoir, London.

Ziegler, A. M., M. T. Gibbs, and M. L. Hulver. 1998. "A mini-atlas of oceanic water masses in the Permian Period." *Proceedings of the Royal Society of Victoria* 110:323–43.

Zolotukhin, V. V., and A. I. Al'Mukhamedov. 1988. "Traps of the Siberian platform," pp. 273–10 in J. D. Macdougall, ed. *Continental Flood Basalts.* Kluwer Academic Publishers, Dordrecht.

Zhou, Z., B. F. Glenister, W. M. Furnish, and C. Spinosa. 1999. "Multi-episodal extinction and ecological differentiation of Permian ammonoids," pp. 195–212 in A. Y. Rozanov and A. A. Shevyrev, eds. *Fossil Cephalopods: Recent Advances in Their Study.* Russian Academy of Sciences, Moscow.

Index